Hudson 1946-1957

The Classic Postwar Years

Richard M. Langworth

To the men and women
of American Motors,
that feisty independent.

Credits

All photographs credited to John A. Conde or American Motors Corporation except the following:

Robert F. Andrews: p.35 (left), p.36, p.37 (upper left, right), p.38, p.39, p.40 (upper right).

Car Life: Dick Adams p.67 (right), p.68 (upper left); Don O'Reilly p.67 (left), p.68 (right); T. Taylor Warren p.68 (lower left).

Joe Weatherly Museum: p.71, p.75.

Art Kibiger: p.15, p.16.

Michael Lamm: p.100, p.101 (left).

Richard M. Langworth: p.58 (upper left, center, right).

Petersen Publishing Company, 1969: p.14.

Michael Sedgwick: p.81 (left).

Special-Interest Autos: p.80, p.81 (right), p.102 (left).

W. W. Tilden: p.70.

This edition published in 1993 by Motorbooks International Publishers & Wholesalers, PO Box 2, 729 Prospect Avenue, Osceola, WI 54020 USA

This is a reissue of the 1977 original edition with corrections and revisions added

Library of Congress Cataloging-in-Publication Data
Langworth, Richard M.
Hudson, 1946-1957 : the classic postwar years / Richard M. Langworth.
p. cm.
Rev. ed. of: Hudson. 1977.
Includes index.
ISBN 0-87938-729-7
1. Hudson automobile—History. I. Langworth, Richard M. Hudson. II. Title.
TL215.H77L35 1993
629.222'0973'09045—dc20 92-33789

On the front cover: The 1949 Hudson Commodore 6 Convertible owned by Bob Hill. *Bud Juneau*

Printed and bound in the United States of America

Foreword

THE POSTWAR STORY of Hudson is mainly one of missed opportunities. Together with Packard, Hudson brought a great reputation, among independents, into the reviving automobile industry. In 1948 it launched the Step-down, one of the most advanced automobiles of the day—but by 1954 it had failed to improve on the basic formula and sales were plunging. In 1951 Hudson introduced the superlative Hornet, which swept the stock car championship for the next three years—but the buying public soon shunned the cars for V-8-powered competitors. In 1953 there arrived the compact Jet—but it was three years after the initial compact car spurt, which had benefited Rambler, and five years before the second great rise of small cars that gave success to the Big Three compacts, Valiant, Falcon and Corvair. Ten years after the war, the Hudson Motor Car Company was no more—a secondary partner in the merger that became American Motors. By 1957 the name had been quietly discontinued, as a result of AMC's drive to dominate the economy car market with the Rambler.

In a way, it had to happen. The field was rapidly narrowing for automotive independents after the war, though it took most of them a long time to realize it. From scores of companies in pre-depression years, their number had shrunk to less than a dozen at Pearl Harbor, and after V-J day only six of these companies returned: Studebaker, Nash, Hudson, Packard, Willys and Crosley. By 1949 the Big Three held close to ninety percent of total sales, a huge lead. One page of national advertising cost them about *ten percent*, per car, of what it cost an independent. In 1953 came the 'Ford Blitz', during which Dearborn tried, and failed, to overhaul General Motors—but the independents, whose dealers couldn't afford to discount cars to match Ford and GM dealers, were in deep trouble. In 1954 Nash merged with Hudson. American Motors, formed out of these two weakening independents and projected to last no more than a decade by many, is with us yet. "Their profit curve goes up and down like a roller coaster," one executive told me, "but they somehow keep hanging in there." And, to a great extent, whatever success AMC has had is due to the same kind of original thinking that marked the Hudson firm for half a century.

This book attempts to investigate Hudson since the war, both as a product and a corporation—to delve into the qualities of the cars that made them great, as well as the factors that weighed against Hudson's survival as a manufacturer. It surprised me that no one had done a book like this already, because of all American companies after World War II, Hudson may have been most consistent in building cars that were exciting, innovative, fun to drive and unmatched in engineering and design safety until years later. The business and economic factors that spelled the doom of Hudson have been somewhat explored in connection with other independents, but I've tried here to relate them to Hudson specifically, by talking to and reading about the men who ran the company and made the history.

It is important to express deep thanks to all of those who spoke with me on the subject, including some who prefer not to be identified as well as some who do. At the head of the list are American Motors President Roy D. Chapin, Jr., former Hudson assistant sales manager; and former Packard President James J. Nance. Both gentlemen took time out from schedules that provide little leeway, and contributed to the accuracy of this book from a business standpoint. Mr. Nance, it turned out, offered information on the four-company merger conceived in the late forties by Nash's George Mason—the first public disclosure of this fascinating story.

It is necessary too, to acknowledge the help of engineers Stuart G. Baits and the late H. M. Northrup, designers Art Kibiger and Bob Andrews, AMC Public Relations vice-president Frank Hedge and assistant PR manager John A. Conde. Bob Andrews and Art Kibiger provided most of the illustrations of Hudson prototypes—difficult items to find these days and the first, to my knowledge, recently published. John Conde not only led me to cavernous sources of material, but read the advance manuscript for accuracy and provided his own impressions of the Hudson Company from his many years in the employ of Nash-Kelvinator and American Motors.

Gratitude is also owed the Hudson-Essex-Terraplane Club, one of the best buys for your money among car clubs today and possessed of some of the most knowledgeable car collectors, including Pete Booz, Bruce Mori-Kubo and Alex Burr. All three helped check the manuscript for accuracy and provided information or ideas which were incorporated into the text.

As ever, the help of Jim Bradley of the National Automobile History Collection, Detroit, and Mary Cattie of the McKean Collection, Free Library of Philadelphia, was indispensible. To Pat Chappell, who rounded up contemporary magazine and periodical references, I am truly appreciative, because without her help we'd probably still be in the research stage. Thanks too to Motorbooks International President Tom Warth and his capable staff, led by Bill Kosfeld and including eagle-eyed copyeditor Elizabeth Berry.

Foreword to the Second Edition

I should like to acknowledge the Hudson-Essex-Terraplane Club, which so graciously received and praised a book on a subject they collectively know much more about than the writer, and who have been so helpful combing out nits over the years.

I thank in particular Dana K. Badertscher, former chief chassis engineer, for taking the trouble to point out several key omissions or errors in the original text and providing the following overdue acknowledgment: "W. J. McAneeny was president and general manager of Hudson during the racing period. His name was omitted from the first edition which was unfortunate. He was a racing enthusiast and without doubt the most influential in promoting our racing program which resulted in our breaking practically all speed records."

As all car books should, this volume contains opinion, both subjective and objective. Massive reiteration of dull facts is only that. I am responsible for the opinions, as I am for the mistakes which have inevitably crept in, and on which I shall be glad to be corrected.

Richard M. Langworth
Putney House
Hopkinton, New Hampshire
April, 1993

Table of Contents

CHAPTER 1
The Postwar Revival
page 6

CHAPTER 2
The Importance of Stepping Down
page 30

CHAPTER 3
The Fabulous Hudson Hornet
page 58

CHAPTER 4
Hudson's Edsel:
The Short, Sad Flight of the Jet
page 78

CHAPTER 5
Italia and X-161:
The Hudsons That Weren't
page 98

CHAPTER 6
The Hash and the End
page 106

CHAPTER 7
Why Hudson Failed
page 124

APPENDIX I
Statistical Summary, 1945-1954
page 127

APPENDIX II
Production Figures, 1946-1957
page 127

APPENDIX III
Hornet Racing Record, 1951-1955
page 129

APPENDIX IV
Literature, 1946-1957
page 132

INDEX
page 134

The Postwar Revival

THE DISTINGUISHED NAME of Hudson, so proudly placed in the American automobile industry for nearly fifty years, ironically stemmed from no brilliant designer or engineering genius, but rather a Detroit department store owner who never became actively involved in building automobiles. J. L. Hudson, one of the company's founders, was tapped mainly to help finance the new car back in 1908. Though a blueprint for that car has recently surfaced, bearing his penciled scrawl, "Let's go with this," it is literally certain that finances were his only contribution aside from his name.

One might imagine, if he didn't know better, that such an obvious salute to the car's 'money fairy' augured ill for its future worth. Couldn't there have been a better name than J. L. Hudson's? Perhaps, but Hudson it was. And as the years went by, that name came to represent much that was good about American automobiles.

If the marque ever found its way onto a questionable product, such instances were rare. Hudson's history was largely a happy one, its cars always full of character—often unusual, sometimes revolutionary, rarely if ever dull. The triangular Hudson logo, adopted in 1909, was later decorated with the coat of arms of Henry Hudson the explorer: ships, to "symbolize the adventuresome and pioneering spirit of bold and imaginative engineers," and fortresses, to represent "...the strength of the organization throughout the world. Placed on the Hudson shield, this 'coat of arms' is a mark of distinction on cars of quality." For once, the ad flacks were right.

Formally, the Hudson Motor Car Company was organized on October 29, 1908, by the aforementioned J. L. Hudson, his relative Roscoe B. Jackson, Howard E. Coffin, F. O. Bezner, Roy D. Chapin, J. J. Brady, Lee Counselman and Hugh Chalmers. Chalmers, who built the Chalmers-Detroit, was said to have provided some parts for early Hudsons, but this rumor has never been substantiated. The real drive was mainly provided by Chapin and Coffin—one smiles at the thought of naming the car after the latter. The two had met at Oldsmobile, where Chapin was sales manager and Coffin chief engineer; according to Chapin they were "great friends, and worked together very closely. From the first, Coffin had a wonderful grasp of both engineering and design. When he worked out something new we talked it over together—both from the engineering and the sales standpoint. . . .Between the two of us we managed to keep pretty closely in touch with everything that was going on in the motor field." The relationship began in 1905, three years before Chapin, Coffin and their associates united to build the first Hudson car.

The four-cylinder Hudson 20 for 1909 began a tradition of individuality by offering selective sliding gear transmission, and selling for a low price of $900, f.o.b. Detroit. It was built in a small, 80,000 square-foot plant, staffed by 500 men, and backed by an initial capitalization (most of it J. L. Hudson's) of $20,000. Over 4,000 model 20's were sold in 1910, Hudson's first full year of manufacture, a record for any new make at the time, and the first sixteen months of production netted close to $4 million in sales.

The Hudson administration building on East Jefferson Avenue, 1951, and Hudson's triangle emblem.

The 20, Hudson ads said, was "good looking, big and racy. Note the graceful and harmonious lines. Observe the sweep of the fenders and the frame." Without much imagination one can visualize the same words being applied to the main subject of this book, the Step-down Hudson of nearly forty years later.

Thus Hudson was on its way. In 1910 a new and much larger plant was purchased "in the country," at Jefferson and Conner Avenues, about five miles from the center of Detroit. At that time the land was still surrounded by its original split rail fence. Hudson erected a $500,000 factory with floor space of 172,000 square feet, and shortly offered the model 33, pioneering 'unit power'—engine, transmission and clutch in one assembly. Good sales soon resulted in a doubling of plant size.

Hudson quickly established a six-cylinder specialty. Its 1913 model 54 was the first medium-priced six to exceed 60 mph; this was followed by the 1914 model 6-40, the first medium-weight six at only 2,680 pounds—well under the ponderous heavies among its contemporaries. Hudson switched over to six-cylinder engines exclusively that year, and also introduced the world's first four-speed overdrive transmission and counter-balanced crankshaft.

For 1916 the company introduced its revolutionary Super-Six, one-upping the industry with the first fully-balanced, mass-produced crankshaft. Until its appearance, production cars lacked balanced crankshafts and were forced to maintain a very low engine rpm, about 1200 to 1400. When balancing was done, it was done by hand—and only on luxury cars. The Super Six, first shown to the press in November of 1915, was a complete shift from this tradition: Its fully balanced crankshaft allowed as much as 3500 rpm, and the engine developed almost twice as much power per cubic inch as any other contemporary car. Featured also was a non-detonating cylinder head, which permitted a then-incredible 5:1

Hudson display at Detroit Public Library in 1946 attests to many claimed company 'firsts.' Though some were not really Hudson innovations, the firm was a consistent leader in automotive development from 1909.

compression ratio despite the low octane fuel of the period. As a result, the Super Six was able to outperform cars with engines more than twice its size.

At Sheepshead Bay raceway in 1916, Ralph Mulford took a perfectly stock Super Six to 102.43 mph over ten miles—a new record. He then broke the twenty-four hour track record by making 1,817 miles in that period. Stripped down Hudson racing cars soon began entering contests all over the country, usually with mildly modified engines in 'speedway' bodies. With Mulford and Ira Vail up, a Super Six won the 'open class' record against all comers including eights and twelves; at Chicago, Mulford broke two American speedway records and averaged 104 mph for 200 miles.

Hudson's new skein of victories was the most impressive by far of any automobile manufacturer, and the public naturally responded enthusiastically. Over 26,000 Super Sixes were sold in 1916, and by the end of the year the Jefferson Avenue plant had reached 847,000 square feet.

Its early rapid growth did not prevent Hudson from making progress in many diverse areas of automotive design. For example, it had a self-starter in 1911—years ahead of many others. In 1913 it displayed a straight-through cowl design flowing right past the dashboard, unlike most of the buggy-styled cars of the day. In 1915 Hudson introduced a convertible sedan, a closed car with removable windows and side posts. And in 1919 it brought forth the four-cylinder Essex, to the loud applause of industry and public alike.

The Essex four-cylinder F-head engine was capable of outstanding performance in its class. Four of them, for example, carried the first automotive 'fast mail': Two cars left San Francisco for New York while two more left New York for San Francisco, each carrying U.S. mail. In spite of bad roads, rain, mud and the usual conditions prevalent in those days on most of America's highways, the average speed for all four cars was four days, twenty-one hours, thirty-two minutes. One of them broke all the transcontinental speed records, and each driver was sworn in as a 'special mail carrier.' The Essex also won class victories at Pikes Peak and captured the Penrose trophy, awarded for the fastest time.

If there was any questionable product in the prewar years, it may have been the 1924 Essex six. The engine was both slower and smaller than the four it replaced, and no one has ever explained Hudson's reasoning in introducing it. It did not stand up to high speeds the way its predecessor had, though it was very smooth and flexible for around-town use. "It makes a fine parade car, even today," says one Hudson collector, "but take it out on the road and you usually had trouble if any sustained speed was attempted."

More significant than the six was the 1922 Essex coach, the industry's first low-cost closed car—a stunning Hudson innovation. Over the next decade, the closed coach would replace the touring model as the dominant body style, and Hudson may claim credit for starting the trend. Essex coaches sold like nickel hamburgers: By 1925, Hudson had moved into third place in the production race behind Ford and Chevrolet. Plant floor space was over two million square feet, production capacity triple that of 1920.

In 1927 Hudson developed a new Super Six with a huge in-line F-head engine, used for three years before the depression assured its demise. Of this new Super, enthusiast Bruce Mori-Kubo says "It could pass anything on the road, with the possible exception of a gas station. As a result it quickly became a favored car for the numerous liquor runners during Prohibition. Actually, the gas mileage wasn't all that bad, though for 1927 it may have left something to be desired. A later 1927 was improved in this respect and used through 1929. This big six really tore up the record book, and some records were still standing only a few years ago—they may still stand today."

Still the innovations came. In 1928 Hudson became the first manufacturer to build its own production bodies in its own plants. Bodies for custom models had been built by such coachworks as Murphy of Pasadena, California and Biddle & Smart of Amesbury, Massachusetts, but when those went out of business Hudson took on this portion of the operation as well. With the exception of open cars through 1934, Hudson built all its bodies until 1953, when it contracted for the production of a new compact by the Murray Company.

With the depression at its lowest ebb, Hudson developed 'unit engineering' in 1932: Bodies and chassis were built in a single, bolted-together unit. The process eliminated hundreds of pounds of weight and enhanced performance: 120 U.S. stock car records fell to Hudson over the next seven years, including the production-car speed record for a twenty-four hour run—87.68 mph, set by a Hudson Eight at Bonneville. It was still the mark to beat a decade-and-a-half later.

By 1936, Hudsons were using all-steel bodies (again ahead of most competition), steering-column gear shifting and rotary equalized brakes. Hydraulic brakes, with mechanical back-up operated by the same foot pedal, were introduced in 1936. This became mandatory in the U.S.—in 1967! The battery-under-the-hood was a Hudson innovation (in the past it had usually been found under the floorboards) as was dual throat carburetion for six-cylinder engines, the dashboard-controlled hood lock, Airfoam seats and Drive-Master transmission, which eliminated the use of clutch and gearshift in all forward speeds.

Unfortunately, all this brilliant independence and originality was little help against the depression: Hudson began slipping along with the rest of the industry immediately after the stock market crash in 1929. Its all-time record earnings of $21.4 million had been scored in 1925; 1929 was a record sales year at $201 million, but that figure plummeted to $78 million a year later while Hudson just managed to remain in the black. Again in 1931 sales dropped, to $38.2 million; for 1932 it was $25.8 million. Losses of approximately $8 million were recorded in both those years, and it began to look like Hudson, its spirit and individuality notwithstanding, would go the way of Peerless, Jordan and countless other victims of the economic malaise. But those who thought so had not counted on Roy Dikeman Chapin.

Chapin was appointed Secretary of Commerce in the last days of the Hoover Administration, temporarily absenting him from Jefferson Avenue. He returned in May of 1933, to face the third successive losing year with Hudson $4.4 million in the red. But a scant eighteen months later he had

Roy Dikeman Chapin (second from right), with (left to right) English consulting engineer Reid Railton, UK works manager Thomas and export manager Al German, as pictured in 1935.

British cars using Hudson components included the Brough Superior and Railton of the thirties.

raised $6 million in loans to keep the company going, which was all Hudson really needed—it already had the product in its new Terraplane.

This brilliant outgrowth of the Essex, introduced in 1932, was a sure seller. It was fast (up to 80 mph) and economical (up to 25 mpg), and it sold for as little as $425. "In the air," the ads said, "it's aeroplaning. On the water, it's hydroplaning. On the ground, hot diggety dog, that's Terraplaning!"

Terraplane it did. Hudson literally tore up the record book with its new wonder car. Chet Miller set an all-time record in a six at Pike's Peak, and again a Hudson-built car took the Penrose trophy. Then Al Miller set a new and even faster record with a Terraplane eight, and Hudson claimed a Penrose monopoly. Within a year of its introduction the Terraplane broke over one hundred stock car records which Hudson would refer to as late as twenty years thereafter—because they were still standing.

Hudson sales had bottomed at $23.5 million in 1933; by 1936 the Terraplane had them up to triple that, and the company made a comfortable $3.3 million profit, the largest in seven years. Sales were aided by the patronage of aviatrix Amelia Earhart, and even John Dillinger pledged loyalty to Hudson—he always used a Terraplane to outdistance the law.

Unfortunately, Roy D. Chapin was not able to withstand the strain of saving his company, along with the Guardian Trust Company, of which he was a director. Suffering from worry and overwork, he contracted pneumonia and died in February 1936, at only fifty-six years of age. But the miracle he had wrought was there for all to see:

Year	Production	Rank	Sales	Profit (Loss)
1930	113,898	4th	$78,094,714	$ 324,656
1931	57,825	7th	38,235,626	(8,523,906)
1932	57,550	4th	25,861,671	(8,459,982)
1933	40,982	7th	23,521,458	(4,409,930)
1934	85,835	5th	52,567,561	(3,239,202)
1935	101,080	8th	63,077,415	584,749
1936	123,266	8th	77,150,680	3,305,616

One ominous factor, despite the apparent turning of the corner in 1935, was the fact that Hudson's market share was declining rapidly. Over 100,000 cars had been good enough for fourth place in the industry in 1930; by 1936, with the rest of Detroit reviving rapidly, it could assure no higher than eighth. And anything less than 100,000 meant a correspondingly lower industry ranking, as was shown when a recession hit the industry in 1938:

Year	Production	Rank	Sales	Profit (Loss)
1937	111,342	8th	$74,502,130	$ 670,716
1938	51,078	8th	38,845,238	(4,670,004)
1939	82,161	9th	58,036,297	(1,356,750)
1940	79,979	12th	60,631,377	(1,507,780)

The management team faced with the problems and implications of this record was led by A. E. Barit, conservative and austere, who had succeeded Chapin as president in 1936. Barit had been with the organization since January 1910, less than six months after the first Model 20 had rolled off the line. His initial duty was as secretary to Bezner, the founder who headed the purchasing division, so Barit was closely involved in the purchasing that was a result of the half-million dollar factory expansion program in 1910. He was appointed assistant purchasing agent in 1911, and relieved Bezner in 1915; secretaryship of the company was added to his duties in 1923. Barit was named vice president and treasurer in 1929 and general manager in 1934; he retained the latter title and functions after becoming president in 1936.

His years in Purchasing gave Barit considerable experience, at least in the accounting end of the business. "He was an outstanding corner cutter," remembers Roy D. Chapin, Jr., board chairman of American Motors and son of Hudson's Chapin, whose first industry experience was with his father's company. "There was not a dime's worth of overhead in that place if there was any humanly possible way to get out of it. Nobody was even allowed to buy a three-ring binder. You'd scrounge around until you found one. Mr. Barit ran it that way. And if the production at the end of a day was three cars off one way or the other, all hell broke loose. If it was supposed to be 320 it was *not* going to be 315 or 326, it was going to be 318 to 322. If the profit and loss estimate was give or take $50,000, it was high, and to tell you the truth I don't know how he did it."

Barit's head of engineering was Stuart G. Baits, age forty-five in 1936. A University of Michigan graduate who had joined Hudson as a draftsman in 1915, Baits served as chief engineer beginning in 1928 and a year later he was made a director. Baits became second vice president and assistant to Barit in 1934 (Engineering was headed by H. M. Northrup), and on Barit's ascension to the presidency, Baits became first vice president. Roy Chapin, Jr., remarked that Baits was "briefly, a genius. A very shy but attractive person once you got to know him. He was a camera nut and seemed always to have one suspended round his neck. He was a man of culture, who enjoyed music and art." In the latter respect Baits and Barit were much alike, but Baits also enjoyed the outdoor life—fishing, hunting, guns and sports, like football. Such pursuits were outside the realm of the President's interests.

Chief of Hudson styling was Frank Spring, an aeronautical engineer in the Army Signal Corps during the first World War who had later designed and built the Paige-Detroit engine in 1919-20. Spring designed a complete car for the Standard Steel Car Company in 1921, including its overhead camshaft straight eight engine. In 1922-23 he was chief engineer for Courier in Sandusky, Ohio, and helped produce their advanced automobile featuring a dry-sump oiling system, one-shot lubrication and four-wheel brakes. Spring's styling experience was received at the Walter M. Murphy coachbuilding firm in California between 1923 and 1931, where he was general manager and consulting engineer; at Murphy, Spring designed custom airplanes as well as automobiles. His customers sent him the cream of American and European chassis—Packard, Peerless, Pierce-Arrow, Auburn, Cord, Duesenberg, Rolls-Royce, Minerva, Lincoln, Marmon, Franklin and, of course, Hudson. Frank Spring designed a body for the famous Doble steam car, as well as the prototype 16-cylinder Peerless. He also designed the interior of the Douglas Dolphin, the first commercial closed aircraft, and that of another plane produced jointly by Lockheed,

A. Edward Barit.

Stuart G. Baits in 1936.

Frank S. Spring in 1931.

Vultee and Northrup. Spring was hired away from Murphy by Hudson in 1931; he was named director of styling, and would retain that position through 1955, thus directly influencing all the postwar cars built at Jefferson Avenue.

Robert F. Andrews, later one of the Studebaker Avanti's designers, was a very junior member of Spring's styling team when he hired on at Hudson in 1944. Providing a vivid description of this many-faceted man who figured so prominently in the marque's postwar history, Andrews called Spring "the swinger of the group. He was quite a guy. Trim and wiry, I discovered him to be a sportsman pilot, an avid sports car buff and a motorcycle enthusiast. One of those big shaft-driven BMW motorcycles was his frequent transportation to and from work. I felt comfortable knowing we would be speaking the same language."

Discussing Spring, this writer suggested to Andrews that it wouldn't be hard to visualize him in an open aircraft, white scarf trailing, on a solo flight across the Atlantic. "Oh he would have done that in a minute," Andrews remarked. "I remember one plane of his, a Beech staggerwing. He flew it constantly, but he was always fiddling with the controls. I recall that once he was flying one of the engineers—Baits or Toncray—and he was doing his usual thing—fidgeting. Suddenly the plane went into a dive. Spring decided it was a May Day. 'Bail out!' he said, and sent his passenger out in a parachute. But he stayed aboard and kept fooling with the controls. Finally he got it right and the plane started to work again, but meanwhile his passenger landed with a broken leg—he had never bailed out of a plane before! Yes, Spring was quite a character."

Frank Spring was killed in an auto accident in 1959, and that's too bad. He would have been a great man to talk to. As Roy Chapin, Jr., told this writer, "Though Hudson had some very competent people who could translate the idea into reality, Frank was the dreamer and always was very much ahead of everybody else. That's really what his role was. We had an interesting assortment of talents at Hudson. Spring would dream an idea up, another fellow would figure out how to make that into a practical piece of equipment, a third guy would figure out how it could be put into production. Between the three of them they were really quite a team. But most of the generation of ideas came out of Spring."

The business situation confronting these leaders of Hudson in the late thirties was one of malaise. With a production capability of at least 200,000 cars, the plant rarely produced fifty percent of that number, and often much less. It was really the war that pulled Hudson out of its bad profit situation. Said *Time* in 1948, "The war gave Hudson another chance. After a 1940 loss of $1.5 million, Hudson netted an average of nearly $2 million a year making antiaircraft guns, 700-cubic-inch six-cylinder invasion barge engines and aircraft parts. By war's end, [Barit] was determined to recapture Hudson's former place as a leader of the independent motormakers."

Barit was in a good position to make that effort. Operating on cost-plus defense contracts, Hudson had scored brisk sales though netted only modest profits:

Year	Sales	Profit
1941	$ 66,827,146	$3,756,418
1942	106,667,466	2,122,020
1943	154,946,558	1,637,958
1944	88,593,265	1,698,634
1945	71,740,601	673,248

Significantly, after V-J Day in August 1945, Hudson ordered $40 million worth of materials, indicating broad manufacturing and distribution plans. The military cooperated by vacating Jefferson Avenue by February 1946. Among all manufacturers of major warplane components, Hudson was one of the first to complete its plant clearance assignment. The pull-out was celebrated on February 13, 1946, with contract termination officer Major W. C. Little praising the cooperation of the civilian and military workforces in bringing about the speedy clearance. Hudson's war record was proudly cheered. The company had built more than half of the ninety-eight foot fuselage section, wings and ailerons of the B-29 Superfortress, plus wings for Curtiss Wright Helldivers and Lockheed P-38 Lightnings, cabins for Bell Airacobra P-63's, fuselage sections for Martin Marauder B-26's and numerous other items—mine anchors, tank parts, guns and ammunition, gas tanks and radar extrusions. Twice during the war Hudson had won the Army and Navy 'E' award for its defense effort.

At war's end, Hudson properties covered seventy-seven acres and offered three million square feet of floor space. As in the past, Hudson was prepared to manufacture an unusual proportion of its own parts: sixty-seven percent of the completed car, in terms of dollar value. An entirely separate plant was set aside for the production of axles and gears, another for engines, yet another for bodies. "This concentration of manufacturing

Final Hudson leaves Jefferson Avenue in February 1942, as company shifts to fulltime war production.

facilities," the company said, "results in a very large output per square foot of floor space and a rapid turnover of inventory." The firm's total plant and equipment investment was something over $50 million, $90 million counting dealerships and distributorships. In addition to its Detroit factories, Hudson had assembly plants in Great Britain and Canada. Though 1946 found none of the original founders still alive, Hudson's executive and production worker ranks included many twenty and thirty year veterans.

Prompt clearance of the defense operation allowed Hudson to start production of cars the same month Japan surrendered: Thirty Super Sixes and a Commodore Six were built in August, and by the end of 1945 the plant had turned out 5,396 units for eleventh place in the abbreviated calendar-year production race. The cars were not unfamiliar. They could be recognized simply as face-lifted versions of the 1942 line.

It could have been expected. Getting back into car production was of paramount importance, Hudson had the prewar dies and the facilities with which to build cars. And the country was crying for automobiles. Still, many revolutionary or at least unique ideas had been considered during the war: If Hudson hadn't put any of them into production just yet, one could not assume the company had been asleep. Anything but.

Front and rear views of Hudson experimental sportster, 1942.

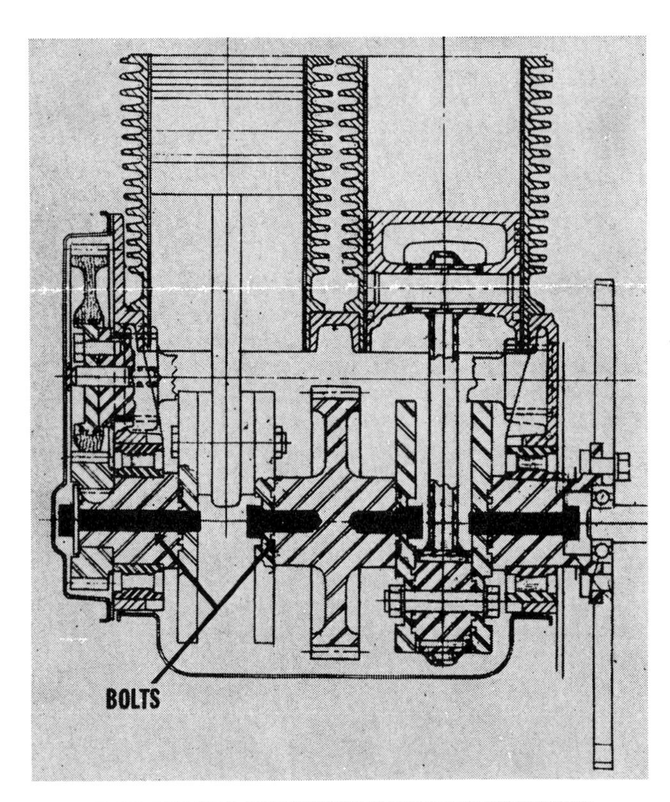

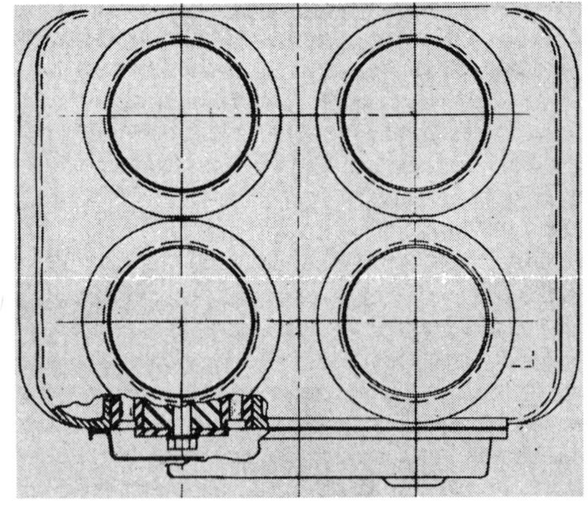

Sportster used 132-cubic-inch square four with built-up crankshaft composed of replaceable, bolt-up throws. The throws were machined instead of cast and could be replaced individually. Twin cranks rotated counter to each other, promoting good balance and smoothness.

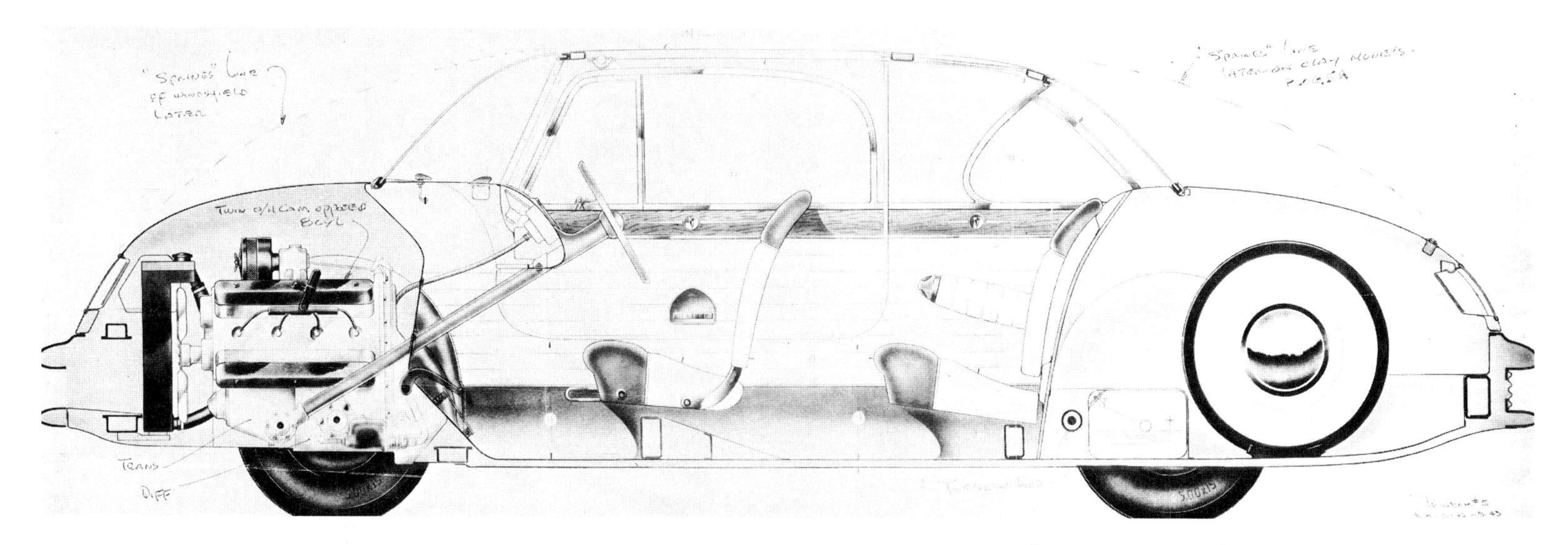

Program 5 followed Sportster in 1943. Three-wheeler used full-perimeter bumpers and had rear-wheel steer.

Sources state that approximately ten significantly different design projects had been considered in a small way while Hudson bent to the war effort. Frank Spring himself had some radical visions: front-wheel drive front-mounted engines, rear-wheel drive rear-mounted engines, aircraft construction techniques, full unit bodies. So did those under him. Roy Chapin, Jr.'s, remark about all ideas coming from Spring notwithstanding, Hudson had some brilliant designers. One of these was Spring's chief lieutenant, Arthur Kibiger, who later became American Motors' styling technical manager.

Bob Andrews calls Kibiger "really the smartest stylist there. He was quiet and not very extroverted, and many people didn't know who he was. But he was the guy who talked about power-to-weight ratios, and the one who knew what the 1100cc Fiat could do. He was constantly pushing power-to-weight, and one of his disenchantments in working on the heavy postwar production designs was that they wouldn't perform as well as our prewar straight eight."

Kibiger, whose future-car ideas actually began to occur before the war, realized after Pearl Harbor that anything would be salable by the time the conflict ended. This situation, he felt, was a custom opportunity to dramatically improve the overall concept of the automobile in one giant leap. By war's end a car-starved public would have to buy—conservatism would not be a governing factor.

One of Kibiger's earliest ideas was an eighty-five-inch wheelbase two-seater that would be only four feet high, along lines mentioned by several other manufacturers. (Henry Kaiser was researching three small cars concurrently, the models 444, 555 and 666. The '444,' for example, stood for $444 price, 444 pounds, 44 horsepower.) Kibiger's car included many popular design ideas of the day such as unit body, hidden headlights, wrapped windshield and curved door-glass, full perimeter bumper—along with some distinctly his own: a molded plastic hinged roof, a mid-mounted aircooled four cylinder engine displacing 132 cubic inches with two opposite-rotating crankshafts.

Kibiger reflected on this and other Hudson prototypes in 1969 with Mike Lamm, then of *Motor Trend*, who reported that a full-sized prototype of the two-seater was actually built—in total secrecy because material was scarce, during WWII days. It eventually grew to a larger wheelbase and acquired a conventional Hudson six-cylinder engine. "Also the perimeter frame and bumpers were over gauge, so the car took on much more weight than intended. The prototype, which was never photographed, ended up with belt drive, and this didn't help performance at all. Finally, the sportster simply died of complications."

Another Kibiger dream car that awoke great expectations on the part of Frank Spring was Program 5, a car Andrews remembers as "the 'bug,' a Skoda-type with wrapped windshields and bumpers, a torpedo-like thing

Four-wheel derivation of Program 5 reached clay stage in 1943. Kibiger proposed standard air conditioning, fixed windows and gullwing doors.

that was of great interest to Hudson for awhile." It was a five- or six-passenger car with torsion-rod-activated gullwing doors, a perimeter frame and Goodyear Torsiolastic suspension which used rubber-jacketed suspension couplings instead of conventional springs—and was later found on the postwar Tucker and Keller Kar. Among the many unique features, its engine had to be the most imaginative: horizontally opposed with double overhead camshafts, its twin cranks were built up, with bearings and crank throw cheeks grooved, ribbed and lapped. Each throw was separately machined and individually bolted to the rest, allowing for easy replacement and more accurate dimensioning than a conventional casting. Thus a customer could specify a four-, six- or eight-cylinder configuration, made up simply by bolting on the requisite crank throws. Beneath the engine was a transaxle, uniting the gearbox and differential. Whether this arrangement created any crank alignment or strength problems isn't certain.

American cars of the forties have often been ridiculed as heralding Detroit's inability to efficiently design autos that could carry five- or six-passenger loads in less that eighteen feet or a 120-inch wheelbase. Yet Kibiger's Program 5, decades removed from the space-laden compacts of the sixties and seventies, carried five or six passengers on a wheelbase of only 102 inches. It measured only 186 inches long, sixty-eight inches wide and fifty inches high, coming close to today's best compact designs. Its outstanding space utilization was better even than Kaiser-Frazer's eighty percent (seat width to overall width), achieved mainly by thin-section hinged doors and curved side-glass. (The gullwing doors disallowed roll-up windows, so Kibiger planned to feature air conditioning as standard equipment.)

Frank Spring, Mike Lamm reported, "was thoroughly taken with Program 5 . . .A. E. Barit ordered an immediate wooden space mock-up built"—which showed remarkable foresight for a man supposed to be so conservative. But "when Purchasing finished projecting the cost of Program 5, they released a report that Hudson could build a much larger car for the same amount of money. And that put the lid on it."

Spring and Kibiger tried other variations of Program 5's packaged power set-up: streamliners with three and four wheels, the former a rear-steer model that was inherently unstable, the latter with a huge wrapped windshield that proved expensive and very distorsional. By

Hudson plants in 1946, as conversion from war industry is being completed.

1943-44, Hudson's advance product thinking had become more conventional, though as Kibiger says, "we had an awfully good time." And not all Kibiger's work was wasted: Many ideas went into the first all-new production Hudson, unveiled in 1947.

By Christmas 1945, A. E. Barit was pointing to new Hudsons dribbling off the assembly line and promising capacity production "as rapidly as materials are available." Said General Sales Manager George H. Pratt, "Hudson's 1946 models bring many improvements and refinements to the postwar market, including an entirely new front end, new exterior styling and completely new interior, embodying luxury, comfort and convenience." Hudson had changed the old 1942's as much as possible without building brand new dies.

Popular features from prewar cars—Double Safe hydraulic brakes with mechanical back-up, Drive-Master, the Super Six and Super Eight engines—were retained. All models were on a 121-inch wheelbase, in the Super and Commodore series. Each was available with a six- or eight-cylinder engine. Production started with sixes (eights followed as soon as reconversion of eight-cylinder motor machinery could be completed) and four-door sedans. These were followed by broughams and club coupes, then three-passenger coupes and finally convertibles. And in addition to the passenger models,

Hudson UK in London, 1946.

The first postwar Hudson comes off the lines in 1945.

Hudson began early production of 128-inch wheelbase pickups "as part of a program to meet urgent requirements for commercial vehicles of three-quarter ton size." In sum, the 1946 Hudson line-up looked like this:

Model Body Style (passengers)	Price	Weight (lb.)
Series 51 Super Six		
sedan, 4dr (6)	$1,555	3,085
brougham (6)	1,511	3,030
club coupe (6)	1,553	3,015
coupe (3)	1,481	2,950
convertible brougham (6)	1,879	3,195
Series 52 Commodore Six		
sedan, 4dr (6)	1,699	3,150
club coupe (6)	1,664	3,185
Series 53 Super Eight		
sedan, 4dr (6)	1,668	3,235
club coupe (6)	1,664	3,185
Series 54 Commodore Eight		
sedan, 4dr (6)	1,774	3,305
club coupe (6)	1,760	3,235
convertible brougham (6)	2,050	3,410
Series 58 Carrier Six		
pickup (3)	1,522	3,500

Styling's biggest assignment on the 1946 program was to change the face of the car—its most important visual aspect. Bob Andrews was assigned this job, with others, but he gives main credit for the new look to Art Kibiger: "In 1946 we went to a die cast grille from a stamping. Behind the horizontal grille bars, it was pitch black. The car was not a style leader, but that front end was the first in America with a black background, and this is very important—the first really modern front end as people visualized it after the war. Art Kibiger in my opinion gets full responsibility for this, and it was a really fresh approach."

The grille was designed to look more massive and lower the impression of the car, so the bars were horizontal and depressed in the middle. A new Hudson triangle emblem was mounted at headlamp level and indirectly lighted, round sealed-beam headlamps were set into ovals carrying the Hudson emblem at the bottom. Kibiger designed the hood ornament, an upright elongated oval set well back on the hood to heighten the steamlined appearance. On Commodores, bumpers were longer than in 1942 and

Hudsons being shipped in 1946.

The final assembly line, 1946.

wrapped around the sides, an influence of wartime prototypes. Bumper guards were of solid construction, and extra ones were set near the ends of the bumpers on Commodores (optional on Supers). Bright metal beltlines ran the length of the cars, allowing two-tone paint jobs to be adapted. For 1946 Hudson offered nine standard colors, two special colors and three two-tones, but reversed a previous practice by using the darker two-tone above the belt, rather than below. Commodores carried a wider body-length beltline, with cast triangle emblems at the lead ends. Hudson concealed its running board for 1946, but retained 1942's front-hinged lockable hood and twenty-five cubic-foot luggage compartment from the prewar models.

The major mechanical variation on '46's was in the choice of engines. The Super Six was an L-head of 3×5-inch bore and stroke for 212 cubic inches and 102 horsepower; the Super Eight used a 3×4½-inch bore and stroke, giving 254 cubic inches and 128 horsepower. The eight provided significantly more urge, though: It added about 100-150 pounds to the car's

The 1946 Super Six four-door sedan. Art Kibiger designed the 1946-47 front end. Bob Andrews says it was predictive of later blacked-out grilles.

weight, and was notably smoother and quieter—as one might expect.

The only other significant mechanical differences were the bigger clutch, radiator and front springs, larger brake lining and higher battery capacity on eight-cylinder models; and the Commodore's tires—6.50×15 against 6.00×16 on the Super. Options on both cars included Drive-Master, overdrive, vacumatic transmission, Weather-Master heating/ventilation system and a Zenith radio with foot-button control for station selection and volume.

Indoors, there was considerably more leeway for improvement and change over 1942, and here was where the Commodore showed its $100-150 price premium over the Super:

Item	Super	Commodore
upholstery	shadow weave	Bedford cord
brake handle	painted	nickel plated
steering column	painted	nickel plated
front-door vent	latch type	crank type
sun visors	one, hinged	two, swiveled
interior hardware	blue	deluxe type
clock	mechanical wind	electric
instrument dials	red on silver	gold on gold
front floor-covering	gray rubber	carpet inserts
window-finish molding	wood grain	chrome
steering wheel	17″ wood grain*	18″ plastic*
sedan robe-hanger	cord	leather

*optional

Among standard Commodore features (optional on the Super) were airfoam seat cushions, bumper extensions, front fender lamps and a dashboard cigar lighter. The Commodore also offered some items unobtainable on the junior series: sedan rear dome light, instrument light rheostat, rear-center sedan armrest, running-board courtesy lights and bright metal bars in the rear windows. While leatherette was used in both models, the Commodore applied it to seat cradles and cushion edges as well as to arm-rests, seatbacks and door panels. Both models featured warning lights for oil pressure and amperes instead of conventional gauges; Hudson had used warning lights since 1932, a pioneer in this respect, though on the 1946 cars they were placed far to the right on the glovebox door, anything but easily visible to the average driver.

Hudson's Carrier, a three-quarter ton pickup, was the larger of two such models offered before the war and one of the largest-capacity pickups ever built. Mounting the Super Six engine on a special 128-inch wheelbase, it ran six-ply 6.50×16 tires, larger brakes and heavier springs, and was available in three body colors. It used the passenger car dash with vinyl upholstery, and standard equipment included twin windshield wipers, thirty-hour clock, rearview mirror and sedan gauges. Its heavy-gauge steel-reinforced load box measured ninety-three by fifty-seven inches with minimal wheel intrusion, and offered over fifty-six cubic feet of capacity, nearly double that of most competing three-quarter ton pickups. At 3,374 units it was the lowest production model, amounting to only 3.6 percent of the 95,000 units built in 1946, and examples today are unfortunately rare. Aside from its limited glass area, it is quite practical by modern standards, and probably one of the best pickups ever designed.

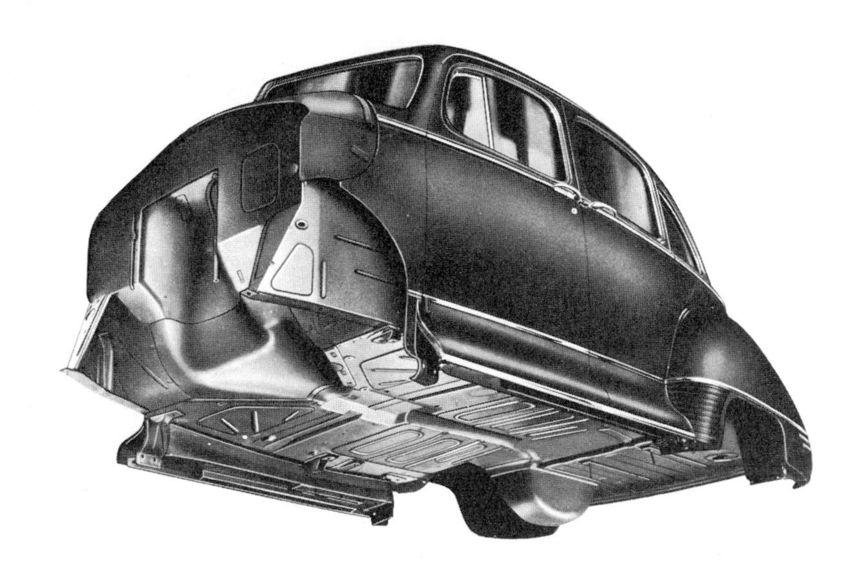

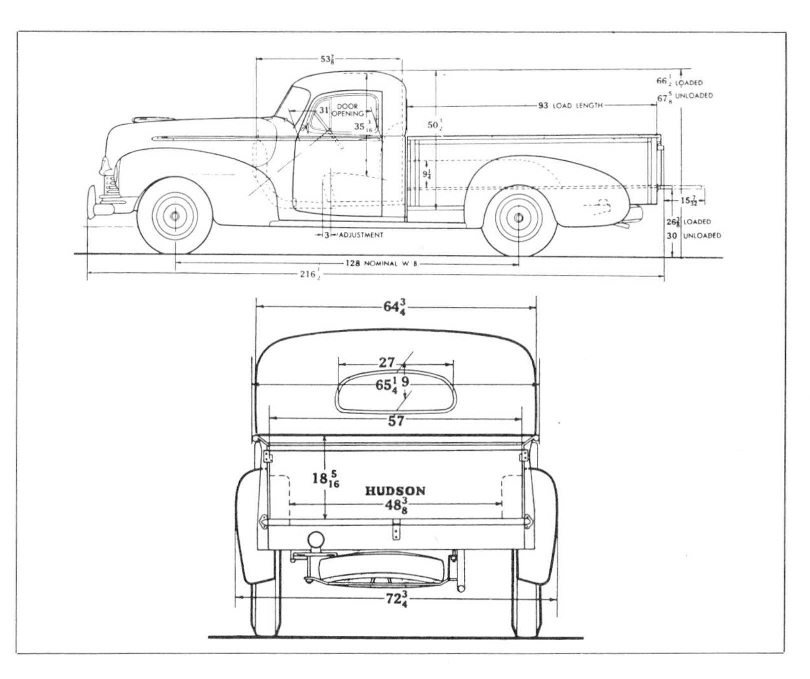

Hudson Monobilt body, 1946.

The 1946 Hudson Carrier pickup.

Pickup's carrying capacity was huge compared to others of this type.

Dealer reaction to the 1946's was said by some to be luke-warm. "At least we had cars," one told this writer. Frank Hedge, now vice president for public relations at American Motors, was with Hudson at the time and remembers that the 1946 model "was regarded as a monstrosity by the dealers, yet it was a beautiful handling car." Still, the '46 was a much better design than most of the warmed-over competition, though not as clean as the '42. In any case the dealer's major problem that year was not its looks, but its numbers. They simply couldn't get enough of them.

Hudson started a second production line, and increased output from 320 to 400 cars a day, on February 19, 1946—just a few days after the last of the military had departed. More cars were hoped for, but officials said, "The plan for continually increasing the rate [beyond 400] will hinge on the same factor of material supplies." There was a shortage of steel, a shortage of coal. There was not enough copper for radiator cores. Most manufacturers had to 'take a strike'—General Motors took one for 119 days—to negotiate peacetime contracts with labor unions who strove to retain the high rates of cost-plus war years. This was coupled with strikes by supplier plants—Hudson had to shut down for five weeks in late March and April 1946 because of one of these.

Gradually, though, production increased. Hudson's rate moved to 480 cars a day on May 13, 1946 and to 560 by the end of the month. By the end of June—when 11,357 vehicles were shipped—the rate was up to 640

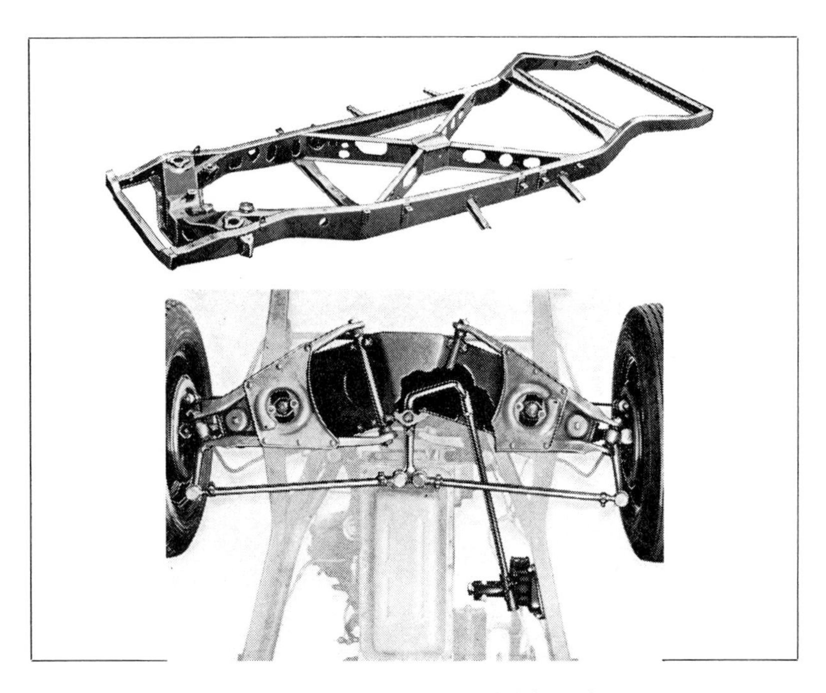

Hudson's postwar X-type frame used extra deep, box-section side rails reinforced with X and cross members. True Center-Point Steering operated from the exact center of the car in order to act equally on either wheel.

The 1946 Super Six convertible.

cars and 17,000 people were employed. More supplier strikes intervened in early August, but when the lines began to move again on August 12, Barit was shooting for "720 cars per day, [which] exceeds the prewar production rate," and he said he'd be hiring 2,000 additional workers.

Hudson ultimately built 90,000-odd cars in calendar 1946, but it wasn't easy. "We have been operating on a hand-to-mouth basis on steel for some time," Barit told a radio interviewer at the end of the year, "and the effect of work stoppages has been far-reaching. While the steel and coal stoppages of last spring were very bad, a company supplying us with two minor trim items, costing but a couple of dollars per car, was closed several weeks because of a strike. As a result, our assembly had to be suspended for six full working days. The loss to the public was over three thousand cars. Our employees lost more than a week's pay. During this past year, we have had an average of some fifteen of our suppliers on strike each month."

According to Barit, the lengths Hudson went to free up a stymied production line were nothing short of heroic. Once, "we were short of a certain type of bolt. . . .Ordinarily we would order these from one of our suppliers, and delivery would be effected promptly in any quantity we desired. The bolt shortage threatened to hold up our entire production. We had to go out and get them somewhere. So each morning we dispatched several of our small pickup trucks to canvass the surrounding countryside for bolts. They scoured cities as far away as a hundred miles, calling on hardware and other stores. Fortunately by getting a few here and a few there they were able to tide us over until our suppliers would take care of us.

"We have a term called 'unbalanced inventories,' meaning that we have an ample supply of some items and a short supply of others. I have mentioned steel as a difficult item. It is out of balance with our supply of certain plentiful items, such as tires. There have been numerous shortages in

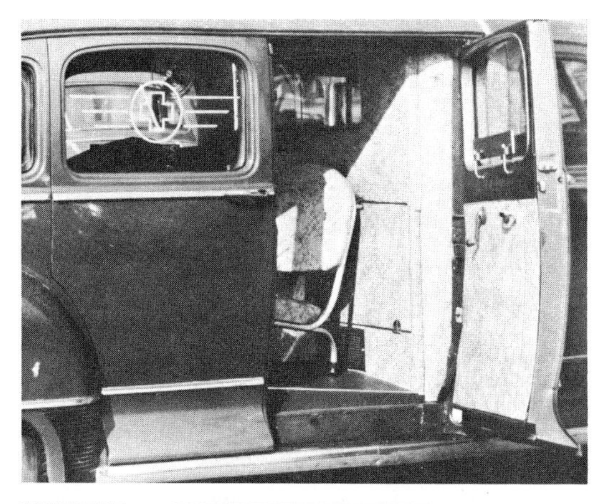

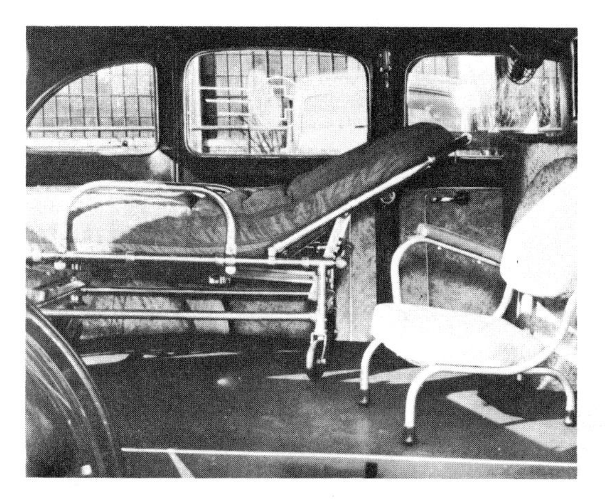

Special bodied, long-wheelbase 1946 Super Six ambulance with detail views.

many other items. Remember there are some 6,000 different parts that go into the manufacture of a Hudson automobile."

Other developments in 1946 were more encouraging. In March the U.S. Office of Price Administration announced that car prices would be advanced to cover wage increases; Hudson raised its prices accordingly, within two days of the order. "We are quite proud of our labor record," Barit said—Hudson had dealt generously with the unions. Also improving was the dealer network, which Hudson had worked hard to maintain during the war. Many new dealers had been appointed since V-J Day, and a lot of them were "putting up new buildings, adding equipment to their service stations or renovating their quarters," Barit said. "In fact, at the present time approximately $22 million has been or is being spent for new buildings. There are many unfilled orders at each of our 3,003 retail outlets, and our job is to fill them. We are also helping our dealers . . .with spare parts for older model Hudsons."

A pleasant development in this year of plant conversion and start-up expense was the 1945 Annual Report. Hudson had made a profit in 1945! A small one, to be sure—$673,248—but a profit nonetheless. "All of the elements necessary to a highly successful operation are present," the report said. "We are now offering a well established product which has been enthusiastically received by our dealers and the public. We enjoy broad facilities for the manufacture of this product, including thoroughly tested special tools and dies and a seasoned organization; we are well financed for all normal requirements; our retail outlets represent in the aggregate a great increase in distributing strength, able to match our capacity; and a well equipped and well organized engineering and research department is constantly at work to insure an advanced product to the buying public."

Hudson made another $146,000 profit during the first quarter of 1946, despite continued production problems, and in May announced a rights issue offering one new share of stock for every seven held by investors. "Because of the current unprecedented demand for automobiles," Barit said, "it is deemed prudent to sell certain shares and to use the proceeds thereof to augment the working capital of the Company." Hudson was girding for the future.

The Commodore Six's grained instrument panel had gold-backed instruments, new Zenith radio with station selection and volume controlled by hand or foot, pushbutton starter and warning lights for oil pressure and amperes. Plastic steering wheel measured 18 inches in diameter.

The 1946 Commodore Six sedan.

From abroad came a variety of good signs. Both the Canadian and British subsidiary plants were profitable. The Stockholm distributor, A. B. Hans Osterman, reported that deliveries of new Hudsons had been made to His Majesty King Gustav and the Crown Prince Olaf. And from Holland, whose queen owned a healthy pile of Hudson stock, came this gem: A citizen of Numegan, anxious to save his 1937 Hudson from the invading

Hudson's policy committee meets in Detroit. Left to right around the table are: Howard Calhoun, Northeast Division; George W. Estaver, Southeast Division; Walter S. Milton, Southwest Division; C. W. Treadwell, advertising staff; N. K. VanDerZee, Assistant General Sales Manager; George H. Pratt, General Sales Manager; M. M. Roberts, Advertising and Merchandising Manager; W. L. Courage, assistant to Mr. Pratt; W. E. Young, Pacific Division; C.A.J. Hadley, Northwest Division; R. N. Hamilton, Jr., Manager Business Management; and E. J. Beguhn, Midwest Division.

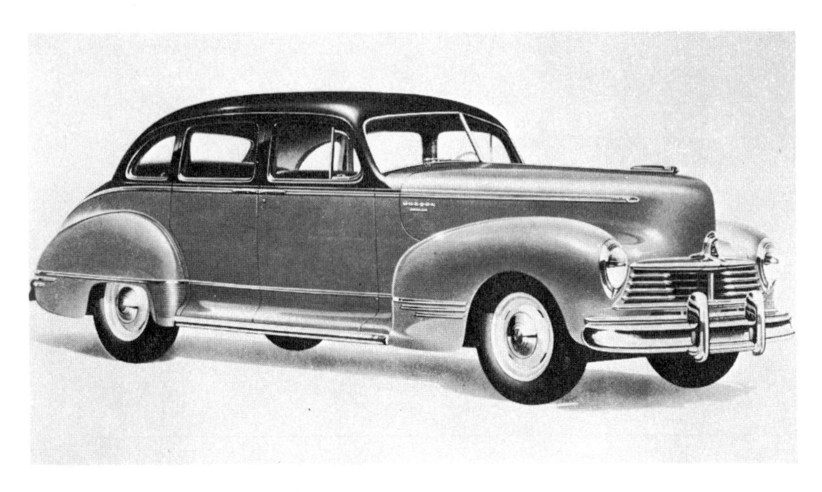

The 1947 Super Six sedan.

Nazis, had buried it in a haystack in 1940, and having just pulled it out and driven 8,000 miles without an incident was mindful of Hudson quality.

Barit had always tried to do right by veterans. He had a special department which trained them in apprenticable trades and placed them in jobs. Former Hudson employees returning from the service were reinstated in their old jobs at Jefferson Avenue. In November 1946, Hudson even developed a special hand-throttle and brake attachment for veterans with leg amputations, and a steering wheel knob with a socket as prosthetic devices for those with hand or arm amputations.

In May 1946, Hudson directors were elected for the ensuing year. The team was familiar. Barit remained president and general manager, Baits first vice president and assistant general manager. H. M. Northrup was vp-Operations, R. W. Jackson vice president, A. Hood treasurer, G. H. Pratt general sales manager and G. G. Behn director-at-large. New additions to the board were investment banker Albert H. Andriesse, Detroit attorney Yates G. Smith and Roy D. Chapin, Jr., son of Hudson's great leader, serving as district manager of the Hudson Sales Corporation.

Production was bound to get better as the labor situation solidified and the supplies of scarce materials eased. It did. Hudson continued to improve its position through 1946, and ended with an income of $2,748,107 for the year, after all charges and taxes, or $1.51 per share of common stock. The dealer body was fast approaching 4,000. In his message to stockholders, Barit reported "a substantial increase in [our] share of industry output, though the number of units [we] built were less than had been hoped. Factory sales for the year totaled 93,880 automobiles. Of these, 90,776 were passenger cars and 3,104 commercial cars. A program costing several millions of dollars has begun to provide for increased manufacturing efficiency and improved research and engineering facilities."

The 1947 Hudsons were almost entirely unchanged from 1946, though the observant buyer would note a new chrome nameplate on the rear decklid, and both right and left door locks. A 70 series model designation replaced the 1946 50 series. Other slight differences included a small lip around the 1947 center emblem housing, making it a little larger than in 1946.

Model Body Style (passengers)	Price	Weight (lb.)
Series 71 Super Six		
sedan, 4dr (6)	$1,749	3,110
brougham (6)	1,704	3,055
club coupe (6)	1,744	3,040
coupe (3)	1,628	2,975
convertible brougham (6)	2,021	3,220
Series 72 Commodore Six		
sedan, 4dr (6)	1,896	3,175
club coupe (6)	1,887	3,090
Series 73 Super Eight		
sedan, 4dr (6)	1,862	3,260
club coupe (6)	1,855	3,210
Series 74 Commodore Eight		
sedan, 4dr (6)	1,972	3,330
club coupe (6)	1,955	3,260
convertible brougham (6)	2,196	3,425
Series 78 Carrier Six		
pickup (3)	1,675	3,000

Once again, the widest model variation was given to the two more 'natural' combinations of Super Six and Commodore Eight, and exactly the same number (95,000) 1947 models were built as 1946's, though the proportions were different. The sixes increased their share from 77,000 to 83,000 and the Commodores increased from 25,000 to 37,000. Again the pickup was a low production model, with only 2,917 produced. Upholstery on 1947 Commodores was herringbone weave, with Bedford cord and genuine leather (standard on all convertibles) or leather-grain vinyl optional. The Super used shadow weave cloth with leather grain optional. Hudsons continued to trumpet the very effective Weather-Master system, in which fresh air entered the car at the cowl, was drained of snow and rain, filtered of dust and insects, heated by a coil if required, then circulated through the car and drawn out from below by a pressure equalizer. Like Nash's Weather-Eye, it was one of the more elaborate and effective heating-ventilation systems of the period.

Hudson hadn't resumed its prewar racing pace as yet, but the factory was proud of a distinguished past: "Sixes and Eights built by Hudson hold 149 official American Automobile Association records for performance and endurance," the ads said—"more than can be claimed by any other stock car in the world. Hudson, for many years, has put its cars through grueling tests—not simply to get its name in the record books but to prove to ourselves and the public what owners may expect of Hudson cars in day-in, day-out service." The records were actually almost all set by 1933. They ranged from one kilometer to more than 20,000 miles in distance and were held for all stock cars regardless of size or class; among these were acceleration and hillclimb marks set against much more expensive competitors. "And they are not confined to performance or endurance—standard models have come through victorious in national economy runs, at normal driving speeds and under normal driving conditions. The thriftiness of a 1947 Hudson will pay you many dividends in fuel costs. And overdrive will better [an] already fine performance and economy."

Although model year production was identical, Hudson had circumvented enough supply problems to build about ten percent more cars in 1947 than 1946, some of these naturally being the new 1948 model. The 1947 total, not counting the Carrier pickup, was 100,393—Hudson's first 100,000 car year since 1937. And on May 28 the company celebrated Hudson number 3,000,000. "Of this total," it announced, "more than 650,000 are in operation today." Still, 1947 was good enough for only thirteenth place in the industry, an unaccustomed low spot. The plain fact was that other manufacturers were solving material bottlenecks better than Hudson. Studebaker, Chrysler, Mercury and the fast-organizing Kaiser-Frazer Corporation all passed Hudson this year; of these only Studebaker had traditionally out-produced Hudson before the war.

It should be noted however that the big manufacturers had plenty of trouble with labor and material. GM, as mentioned, experienced an enormous strike after the war, and its proportion of automotive steel

The 1947 pickup.

Caravan Top option for Hudson pickup was light but rugged, with aluminum frame, made of heavy duck. This is very possibly the vehicle that scoured the country looking for bolts.

Kansas dealer showed the flag, down to Hudson-badged motor scooter.

allocations after V-J Day was lower than before Pearl Harbor. The Big Three were also building less cars percentage-wise—eighty-four percent for the first six months of 1947, against over ninety percent in 1939-41:

Company	1939-41 Share	1947 Share
General Motors	46.2%	42.2%
Chrysler	24.0	21.7
Ford	20.0	19.7
Studebaker	3.0	3.0
Hudson	2.0	3.0
Nash	2.0	3.0
Packard	2.0	1.3
Kaiser-Frazer	——	2.2

There is no easy explanation for the dip in the registration percentage of the Big Three. "Steel shortages lie at the root of each company's production troubles (except Kaiser-Frazer, whose limitation is components, particularly engines)," said *Business Week* in 1947. "But the interesting and somewhat baffling fact is that the largest manufacturers are apparently unable to use their buying leverage to get at least a proportionate share of available steel. One explanation may be that [in 1941] the major manufacturers cut down their production in greater proportion than did the smaller manufacturers, to make room for war material manufacturing [And] the steel industry practices historical allocation."

This didn't guarantee a bright picture for Hudson in the long run. Its postwar working capital was $27 million—compared to the working capital of General Motors, *before* the war, of $600 million. In 1946 *Fortune* called GM "the most fearsomely competent industrial colossus of all times. . .the prime example of the fact that the auto industry has so far exceeded the law of diminishing returns. Because it owns so many of its suppliers, GM probably makes more money selling itself parts than it makes by putting those parts together. So the more cars it makes, the more it can spend on engineering research and sales effort and managerial talent; and the more it can make and sell. As sales increase in arithmetical ratio, efficiency seems to increase in geometric ratio."

This writer has always held that the problems of the independents after the war were, to a large extent, created by the times. Hudson had $27 million, and that looked big. But Chevrolet alone—just one division of GM—had set aside four times that much just to build a bigger plant. GM's market share in 1947 was misleading. Despite wicked strikes and material problems, it would only be a matter of time before the titan started rolling. One of the few independents to see this right after the war was Kaiser-Frazer, in the person of treasurer Hichman Price, Jr., who originally signed for that company's Willow Run plant—even though it was monstrous, much larger than they would initially need. "I was young and I was brash," he told the writer, "and I had a whole lot of ideas. One of them was that in the automobile business—although this had not been proven at that stage at all—the big ones got bigger and the little ones went out of business. My influence, such as it was, was that no we can't [do things on a small scale]: it won't work. We may have a period of three or four years—I remember putting 1950 as the terminal date—in which we can sell everything we can make, and hopefully we can price the things at a level where we can make a good profit. But that isn't going to be enough because it isn't enough volume and it isn't enough business really, in this industry. That was Hudson's experience ultimately, and I was sure it would happen to us. Actually it happened to us earlier than it did to Hudson, and for different reasons. . . ."

The 1947 Commodore Eight convertible.

The 1947 Commodore Eight sedan.

To Price, and a handful of other far-sighted people, Hudson was one of the littlest little ones.

It is literally certain that A.E. Barit had no such forbodings as 1947 opened with Hudson selling everything on wheels. He could take encouragement from decreased independent competition: dozens of prewar companies were no more. Graham-Paige and Hupp were not factors after the war. Crosley was a negligible factor and Willys-Overland was not a direct competitor with its specialty product, the Jeep.

But Barit's greatest encouragement came from something Hudson had wrought all by itself: a brand new, entirely redesigned and breathtaking—for 1947—automobile; a car that received one of the most emotionally moving dealer receptions in the history of the industry. Frank Hedge remembers the day it was introduced to the dealers, in the ballroom of the Astor Hotel in New York City: "The dealers stood on chairs and cheered. I was with a bunch of New York dealers who were alternately laughing and crying. That's what it meant to them—such a tremendous thing. Mr. Barit was just flabbergasted—moved, emotionally. He loved the car, but he couldn't believe the reaction."

The reaction was everything Hedge says it was, and it is well-remembered by dealers who were there. "You've got to understand the temper of the times," one told the writer recently. "Studebaker was already out with a new car and we knew the Big Three wouldn't be far behind. Kaiser-Frazer was not so much a worry, but GM and Ford and Chrysler were. Here we were, trying to peddle the old style cars in '46 and '47, and wondering how long demand would last—and to be fair, many recognized it wouldn't last forever. And suddenly there it was at the Astor—long and low and built like a rock, and sleek—for those days unbelievably sleek. Is it any wonder we threw our hats in the air? Wouldn't you?"

Probably. In 1948 the new Hudson was a sensation, and it would carry the company to the heights of success in the ensuing two or three years. It came, as before, in a Super and Commodore series, though on a longer, 124-inch wheelbase. It came in all the usual models, even a convertible, and with a little waiting you could order a coupe for as little as $2,069. It was, as my dealer friend summarized, "one helluvan automobile."

Hudson called it the Step-down.

CHAPTER 2

The Importance of Stepping Down

"MR. BARIT," AN INTERVIEWER asked early in 1947, "do you have the so-called 'dream car' that artists have been putting on their drawing boards?" Hudson's president was forthright: "Yes. It is our business to have one. We aim to keep abreast of the desires of the public. [But] all that can be said at this time is that the emphasis is being put on increased comfort and safety in driving . . .on glamour . . .and on even better performance." That about covered all the bases—but Barit wasn't exaggerating.

By early 1947, notable changes were evident everywhere at Hudson. A coterie of experts had been summoned to plan and develop entirely new manufacturing and assembly procedures; equipment and major alterations were also required. An entirely new production line—with new conveyors, fixtures and other equipment—was installed, as Hudson's manufacturing engineers took a clean-slate approach to the problem of building cars. A total of 164 new machines were purchased and installed, many of them special-purpose devices built to do a particular job. Close to 1,000 machines and presses were shifted around. At the Jefferson plant alone, 2,615 new tools were designed and built, 13½ miles of conveyor track installed. Traditional assembly line techniques were replaced by new and sometimes revolutionary production processes and a completely new assembly sequence. Said a contemporary brochure, Hudson "utilized every bit of imagination and initiative in working out new fabrication techniques . . .due to the fact that no automobile manufacturer had ever built a car like this before, the most extensive plant and equipment change in Hudson history was necessary."

The result, after the expenditure of about $16 million, was the Step-down Hudson—the first all-new postwar car from the distinguished old company.

We have already seen how Spring, Kibiger and their team progressed from design to radical design during the war; and how Hudson nevertheless presssed into 1945 production what were necessarily prewar cars, to satisfy the overwhelming postwar demand for automobiles. The Step-down evolution formally started in 1943 according to Art Kibiger, after the end of Program 5 and its offshoots. According to Hudson, the Step-down began after management laid down "a set of seemingly impossible demands." Allowing for normal ad flack exaggeration, one might more accurately ascribe to it the words of Ed Barit: "It marked a period of fundamental change in automobile design.

"Emphasizing more comfort and general satisfaction for passengers as well as improved mechanical performance," Barit continued, "we determined to combine low center of gravity, more efficient use of interior space, improved riding qualities and a more rugged basic structure. In keeping with the best Hudson tradition, we called for still another 'must'—extraordinary performance. Previously, one or more of these qualities had

Where tomorrow's motor car will show up soon
This page, an adaptation of a two-page, full-color Hudson message appearing in the Saturday Evening Post for December 6 and Collier's for November 29, is typical of the support Hudson is putting behind its dealers and behind a great new car.
Where tomorrow's motor car will show up soon
THIS TIME IT'S HUDSON
This page is an adaptation of a two-page, full-color advertisement appearing in Collier's for November 29, and the Saturday Evening Post for December 6.
Hudson
HUDSON MOTOR CAR COMPANY • DETROIT 14 MICHIGAN
THIS TIME IT'S HUDSON
DISTRIBUTORS AND PARTS DEPOTS
AUTHORIZED DEALERS
HUDSON
Hudson
HUDSON MOTOR CAR COMPANY
DETROIT 14 MICHIGAN

The 1948 unit construction design used a box-section frame built into the body and encircling the rear wheels; the whole was welded together. It provided girder protection around passengers, recessed floor, lower roofline and more headroom in the passenger compartment.

been sacrificed in the making of automobiles in order to achieve or emphasize others.

"Our first decision was that the car would have a low center of gravity, because it affords the opportunity for great beauty of line and greater safety. Secondly, we determined upon a more efficient use of interior space—to utilize, through design, a greater percentage of the available space." The third demand concerned riding qualities, and "it was recognized that to gain the ultimate in this direction it was necessary to cradle the passengers between the axles.

"Our fourth decision was to provide a more rugged basic structure. While a strong, rugged structure of this kind helps to minimize rattles, squeaks and other annoyances, its benefits go far beyond this. For instance, it has a profound effect upon that very important element we call 'roadability.' Our accomplishment toward roadability far exceeded our expectations. This matter cannot be determined in advance through mathematical computation, but it does respond to fundamental treatment and is looked upon as a reward for sound basic engineering. When the first car was put on the road, we were surprised at the unusually pleasant ride afforded by the combination of low center of gravity, rugged basic structure and the cradling of the passengers between the wheels."

In the last statement lay an important key to Step-down design. Many of Kibiger's wartime ideas employed this basic seating principle: Passengers rode between, but not under, the frame. In effect, the floor was attached to the bottom of the frame rails rather than the top. Though this seemingly wasn't a major accomplishment, it was indeed a forward-looking arrangement for 1947—and it did wonders for the stylists.

Hudson was not a unit-body pioneer, but it had maintained an interest in all-steel and unit-type bodies for many years. Hudson bodies were nearly all steel by 1928 (except for the top roof panel), and fully so by the mid-thirties. Though Nash and Lincoln Zephyr had long preceded Hudson with unit bodies, the firm had evolved the patented Monobilt compromise, in which the frame and body were fastened together without sill plates and with rivets instead of bolts, providing good strength without overwhelming weight. At least one prototype unit body car was made prior to American entry into the war—the car which reportedly languished on the roof of the Jefferson plant until Barit, initially reluctant, "brought it down" after V-J day for another look. Some rumors have it that lower management brought the car down and convinced Barit to drive it home, which sold him. According to Kibiger and several others, this prototype exercised less influence than it has been credited with. Many idea cars ended up on the roof in those days. The formal decision to build what became the Step-down Hudson, using full-unit construction, had been made earlier.

The first production unit body was that of the 1922 Lancia Lambda, though the formulation acquired many early exponents and saw American production in the thirties: Chrysler used it on the Airflow, Ford on the Zephyr. It was mainly inspired by the borning aircraft industry, which had early demonstrated that a lot of weight and material could be saved if a body were designed to distribute stress over its entire surface instead of on a separately constructed steel underframe. On cars, the idea had some weaknesses. It was literally impossible in 1945 (though Chrysler did it later with computers) to predict in advance the structural stress distribution. One either found it satisfactory after the unit was built, or he didn't—and started over. If the stress wasn't properly distributed, reinforcement was required—patchwork engineering which began to put back the weight the unit body was designed to eliminate. There was also a strong tendency for the units to rust—though this was never a serious problem with Hudson. Rust, naturally, could destroy the structural integrity of the unitized car, rather than just the body on a conventional arrangement. Finally, and perhaps more sigificantly for Hudson than anyone else after the war, die costs for unit structures made face-lifts of any serious degree relatively difficult. This threw a block in the way of the annual face-lift game, played so prosperously by denizens of the Motor City.

Hudson's decision, according to one former executive, was to accept the risks in view of the benefits. "The unit body was partly adopted to cut the expense of the dies for the heavy separate frames. Hudson, like many others, was trying to get away from that. I thought it was a mistake, personally. They couldn't easily update the styling, and in Detroit, this spelled trouble."

Stuart G. Baits, inevitably a strong proponent of the unit body, told this writer that "Unit construction is the only logical way to build a car. Lincoln and Nash were not influences [on Hudson]: neither were [they] imitated." But there is some divergence of opinion on how expensive the Hudson body really was. Mr. Baits's impression is that it was "no more expensive to build than the separate frame and body." Roy D. Chapin, Jr., while granting that the Step-down "was an extremely good automobile," feels it was "almost too good because it was expensive to build. It was built in sort of compartments in a sense, which made it very difficult to modify. The structure made any deviation from it very costly. In that sense it had a disadvantage, because it was hard to change to meet styling trends."

Apparently some of the smaller features were not very cost-effective either. Chapin recalls a 1948 Hudson which had been purchased for testing by General Motors (as was that company's wont when interesting new rivals

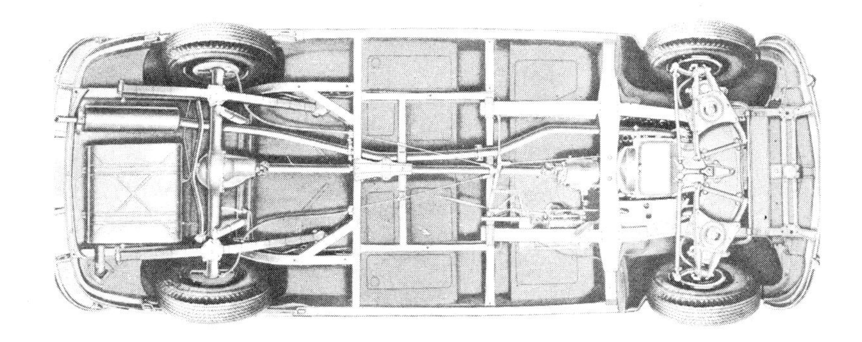

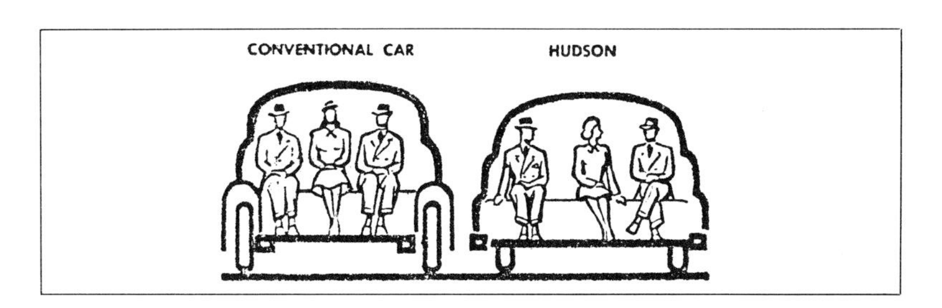

Owners manual shows view from below—the solid passenger protection of the new unit-body frame.

Since the Hudson's floor was under (instead of over) the frame, more than adequate headroom was provided despite an overall height of only 60 inches. Design also permitted good utilization of the car's width for seating.

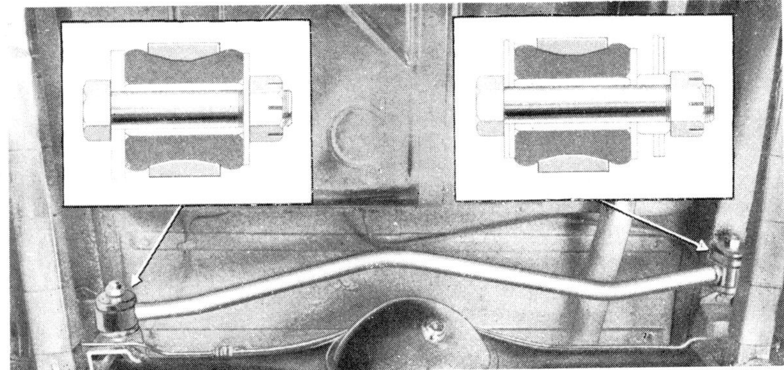

Hudson front and rear stabilizer bars aided roadability.

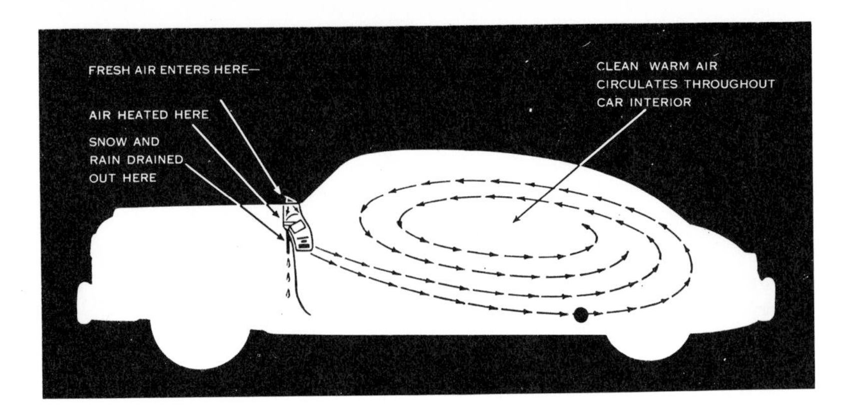

Diagram showing '48 Hudson's excellent airflow-ventilation characteristics.

appeared on the scene). "They used to bring it in to a dealership in the city, when I was assistant zone manager in Detroit. I knew the dealer pretty well, and one day he said he wanted to show me something he'd found in the glove box. It was the notes of the GM executives who had driven it. Essentially they were pointing out that Hudson had such and such a device, 'which we consider too expensive to put in a Cadillac.' There were several things delineated like that which, cost-wise, GM had looked at and discarded. I saw the notebooks, so it's not just a story people make up." Mr. Baits, queried for an opinion on the GM notebooks, had no comment.

Stuart Baits gives credit for the Step-down design to chief engineer Millard H. Toncray. Under Toncray was chief chassis engineer Dana Badertscher, while H. M. Northrup, who had originally preceded Toncray, was vice president for operations. Undoubtedly, Toncray's design made for a resolutely strong, low-slung, rattle-free automobile, and therefore a striking contrast to anything else around in 1948. The body and frame were unitized by welds, the floorboards lowered to the bottom of the frame members. Helping to achieve the lowest center of gravity in the American industry at the time, the main frame members passed *outside* the rear wheels, affording spacious footwells for the rear seat passengers—the rear floorboards were attached to the inner, smaller frame members. Jack pads were built into the frame at each corner behind the bumpers, allowing no-slip jacking (sans the deadly instability of contemporary bumper jacks).

To provide a corresponding driveline, Hudson engineering adopted a two-piece driveshaft which, being jointed, made for a very low floor 'hump.' Side windows, while not curved as per Kibiger's wartime dream cars, did drop into the doors at a seventeen degree angle from the perpendicular, allowing inner door panels to be hollowed out to provide two extra inches of elbowroom. The Step-down offered 61¾ inches of seat width up front and sixty-three in the rear—better than eighty-two percent utilization of its seventy-seven-inch width, the highest percentage in the industry. The Buick Super, for example, had a rear seat only fifty-one inches wide yet was identical in body width (sixty-six percent utilization); the Packard Eight offered a 50½ inch seat against a 77½ inch width (sixty-five percent).

A prevailing, and ancient, Detroit rule holds that a company must not redesign bodies the same year it redesigns engines. But Hudson didn't always follow the rules, and in 1948 it brought out a new engine as well as a new car: the 270-cubic-inch Super Six with 121 horsepower and 3⅞ by

Four-door convertible brougham was contemplated but not built.

4⅜ bore and stroke—considerably squarer than the older six. The eight cylinder unit remained the tried and true 128 horsepower model. Transmission options included overdrive ($101), Drive-Master ($112) or Vacumotive Drive ($47) as in the past. Vacumotive automatically engaged and disengaged the clutch, while Drive-Master handled both clutch and gearshift functions.

If any single physical quality of the Step-down could be singled out it would probably be its lowness. The new Hudson stood only 60⅜ inches high—against close to sixty-seven inches for the contemporary Buick, for example, sixty-six inches for DeSoto and Chrysler, sixty-four inches for Packard, Mercury, Kaiser and Frazer. Even Studebaker, the only prewar company with a new postwar car at the Step-down's introduction, measured close to two inches higher than the new Commodores and Supers. Bob Andrews notes that "this was very inspiring to the design team—we couldn't have developed the designs we did on a more conventional, taller chassis."

Andrews provides a good description of the Hudson styling department during the conceptual stages of the Step-down model. "It was very primitive compared to the studios now, of course, and the full-size work was done in clay—but an awful lot of it was in quarter-size. We had only one full-size buck for clay, and another for front ends. That was it—that was the studio.

"Art Kibiger was chief designer. Spring was director of design, and over him were Messrs. Baits and Toncray. Toncray we saw day to day, but we saw Baits very seldom. In earlier years I think Spring had lost some of his influence over management—it was very engineering-oriented, similar in fact to Chrysler.

"We had only one full-size modeling buck, and those models were usually done in plaster. There was a team of plaster workers who would keep the full-size model up to date as styling changes occurred—and you can imagine the lack of enthusiasm for changes in solid plaster. The plaster workers were mostly Italian—they didn't speak much English and they'd

Recessed door panels provided two extra inches of elbow room on each side. (A 1950 model is shown.)

fight like hell with each other, yet they were really very good friends. But every so often we'd have an explosion. We also had a wood shop where the mahogany models were made for die work—one thing we really proved out at Hudson was die work—staffed mainly by old Germans. You hardly ever heard them speak. They'd just come in, work all day long. They were just marvelous people, tremendous craftsmen."

Art Kibiger says that his staff included, aside from Andrews, a number of people who later won considerable fame: Strother McMinn, now a design instructor; Dick Caleal and Bob Koto, whose work largely resulted in the 1949 Ford; Art Fitzpatrick, later an associate of Dutch Darrin and a contributor to Pontiac. Kibiger also mentions Arnold Yonkers, Arthur Michael and Bill Kirby. "My staff's efforts," he says, "were concentrated mostly on designs of grilles, lamps, bumpers, wheel covers and interiors."

Andrews knew the last three persons well: "Yonkers was a fine, tall Dutchman, an excellent designer and a fine artist, who was fearless in his convictions, spiritually strong, and in addition to designing was a part-time minister. Arthur Michael was one of the most sensitive modelers I was ever

Quarter-inch-scale model from 1944 featured slab sides, Buick-like two-toning on hood, and rear end styling predictive of the 1949 Nash.

Quarter-scale model nears production form for 1948.

Early scale model made approximately 1942 or 1943, invoked envelope body, huge glass area.

to meet. He was a handsome German, kind, and a real pro with a wide classical training as a sculptor. From him I learned a great deal about automotive sculpturing to its highest refinement. It was always a shock to see how the draftsmen would distort and ruin the sensitive areas with their drafting methods of surface development."

Shortly after Andrews joined Hudson, Kibiger added designer Bill Kirby to the staff. "If the word had existed then," Andrews remembers, "Kirby would have been a hippie. He was probably in his middle thirties, already balding, but what hair he had he wore long and shaggy. He wore his suits until they fell apart, a rope invariably served as his belt. . .Bill Kirby was a man of ideas, not of things. A proud member of the 'street people' and self-taught in all areas of the arts, he was incredibly imaginative. He was probably the best far-out designer I have ever known, with a great sense of history and an unbelievable ability to project into the future. . .. After Kibiger and I left Hudson, Kirby was offered the position of chief designer, starting with the stipulation that he dress a little more conventionally. He was incensed, he sat there and said, 'To think that *you* would think that *I* would want to be one of *you*.' The Establishment, with its wall-to-wall ego, obviously wasn't for him. He left the company immediately, never to return. I've heard he didn't even pick up his last check, and when it was mailed to him he immediately sent it back.

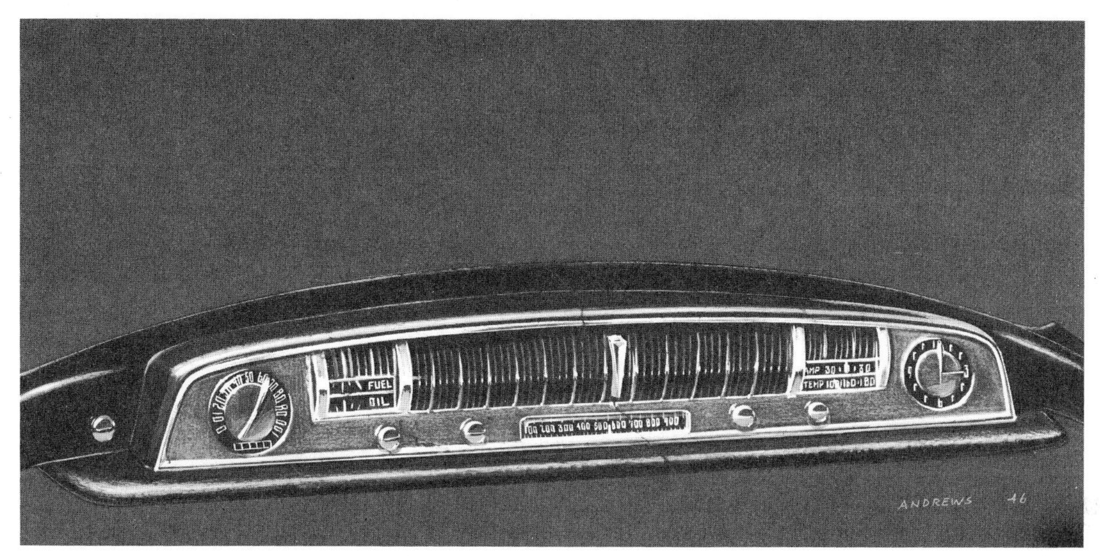

Styling details sketched during mid-1946 included symmetrical dash with huge speaker or ventilation grid and flanking instruments, also 'shadow box' rear license plate holder and illuminated rear deck lift.

The 1941-42 Buick was a kind of benchmark for Frank Spring's styling team.

"There were the people with whom I would work most closely in designing Hudson's first new postwar car. . . .It was readily apparent to me on the day I arrived that everyone at Hudson was most anxious to get on with it."

Art Kibiger comments that the basic shape of the Step-down evolved in 1943, and that after that "only front and rear end shape studies were made. Our first vice president, Stuart Baits, pointed to the full-size model and said, 'I want the windshield to start here and the rear window to start here,' pointing with his fingers. The 1941-1942 Buick provided the high beltline and shallow windows which we did not favor. But only time proved us right."

The Buick, of all things, was treated as a kind of benchmark by the debonair Frank Spring. Andrews and others remember him as a great salesman—a Raymond Loewy-like personality able to sell recalcitrant executives on radical or untried ideas. Somehow, Spring had convinced Hudson that the prewar Buick—a rather handsome car, but nothing exceptional—was the car Hudson had to 'beat' with the Step-down—in 1948! "I never will forget the day," Andrews recalls, "when we were almost done with the final production model. Frank Spring came in and said, 'Well boys, this is it—today we've officially surpassed the Buick!' He was elated—he really believed this, it seemed. He walked out and when he

Some ideas that didn't make it on the Step-down: one-piece windshield, high-riding grille, nerf-type front bumpers and more sculptured flanks.

left, of course, we howled. They took the Buick *very* seriously.''

Probably too seriously. The new Hudson's high beltline was still up-to-date in 1948, but it began to get more and more tired-looking as time went on. Interestingly, Hudson always proudly quoted the square-inch area of its windshield and rear light, but never gave its overall glass area in comparison to other makes.

The basic shape they had to deal with was ''a bar of soap with a bubble on it''—a kind of popular concept foisted off by such as steel, aluminum and other raw material manufacturers as the ''shape of the future.'' But Hudson allowed its designers considerable scope: Phaetons, four-door convertibles, and a sun-roof model were all considered. There was even some interest in a 'pontoon fender' design with definite 'shoulders' (rather like Brooks Stevens' 1962 Studebaker GT Hawk) which some stylists liked. But that was considered too radical and was dropped.

''There was another idea I hated to see abandoned,'' Andrews continued: ''the high front in which we incorporated the main grille area between the headlights, with the headlights carried high. It was a frontal expression not unlike that of the contemporary [early '70's] Pontiac Firebird. But at that time it would have required great courage to introduce. Our high front went up into the finished model (but not the die model); then a decision had to be made, and it was eliminated. Management figured it would be going too far at that time, not a good decision as it turned out— the high front was useful for cooling the hottest section of the radiator core. A high opening would have eliminated one of the few trouble areas of the Step-down Hudson.''

Gradually the car moved toward its final form. Complicated front ends, with multi-bar grilles and double bar bumpers, were eliminated in favor of a grille reminiscent of the 1946-48 models—for family resemblance, probably— and a standard type of bumper. Stylists had thought about rubber inserts for bumper guards, but the idea was dropped due to cost. A rakish, forward-leaning front end was also dropped in favor of a more conventional 'dome' hood, though the basic side elevation remained about the same. Andrews recalls that one of his contributions as a junior stylist was a mid-height creaseline, which helped give character and strength to the side styling; it also served as a two-tone paint divider. Not too many changes were made at the rear, where designers planned triangular taillights repeating the basic Hudson logo.

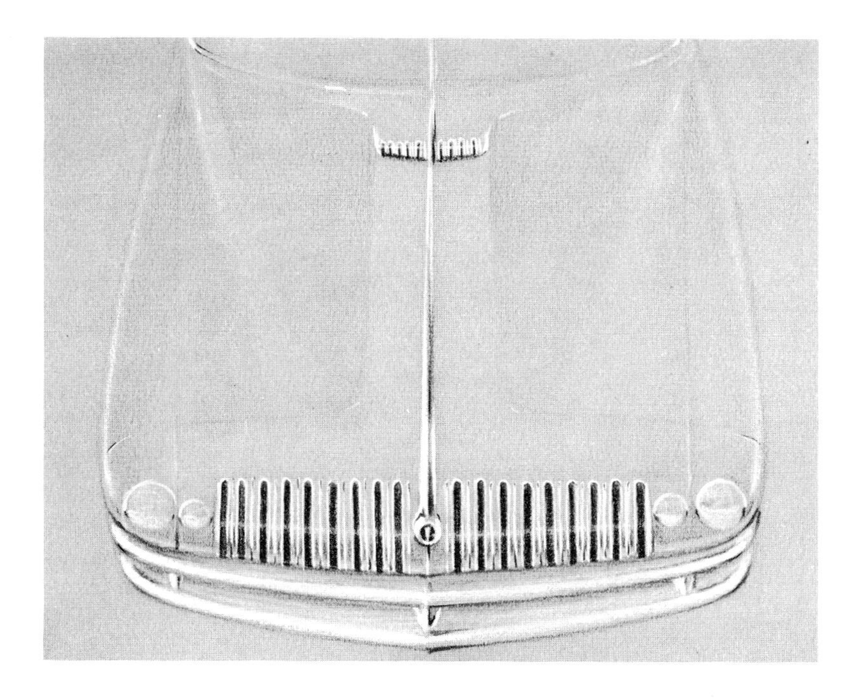

Original sketches made by Bob Andrews, indicate styling thinking for future Step-down. Egg-crate grille motif is from mid-1945, vertical grille (reminiscent of 1940 LaSalle) from early 1946.

Nearing production form, designers were hoping to use a split grille not unlike the Pontiac Firebird.

Another late idea was the high-mounted grille; also shown are the split bumper and Hudson triangle motif on bumpers and taillights.

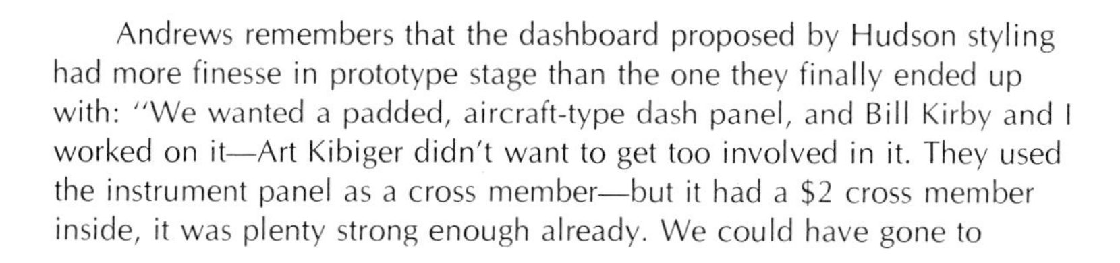

Andrews remembers that the dashboard proposed by Hudson styling had more finesse in prototype stage than the one they finally ended up with: "We wanted a padded, aircraft-type dash panel, and Bill Kirby and I worked on it—Art Kibiger didn't want to get too involved in it. They used the instrument panel as a cross member—but it had a $2 cross member inside, it was plenty strong enough already. We could have gone to

Production prototype clay model on Hudson's factory roof, in clear view from the neighboring Chrysler building! (Unfortunately Hudson destroyed most photographic evidence of prototypes.)

Alterations on the locked-up basic Step-down shape were horizontal taillights, inverted triangle bumper-guards and deck emblem.

aluminum on it." The dash, then, became an angular block with the various controls and instruments tacked on. Of course there were warning lights, instead of gauges—a traditional Hudson feature—for oil pressure and amperes. "I don't think Art liked the idiot lights," Andrews says, "but Frank Spring would alter a whole program to get in a novel feature."

By the spring of 1946 the new car was close to its production form. That point reached, someone in management gave an order, and the stylists were shocked one day to find their quarter-scale plaster models being systematically destroyed by two workers with sledge hammers. "Hudson had many drawings of everything they did long before the war," Andrews told the writer, "and I was almost stupidly honest, because I didn't take any home—I'm relatively sure they threw them all out. . . .Art Kibiger was angry over it, but we never could find out who had given the order. Shortly afterward Art left to become chief stylist at Willys-Overland, in the program that eventually culminated in the Aero-Willys. Six weeks later, in the spring of 1947, he telephoned me and offered me a job, and I left Hudson."

The first glimpse the two had of their handiwork was when the production Step-down was displayed at the Masonic Temple in Detroit during its December 1947 introduction. With Kibiger, Andrews came up from Toledo to see it, "purely as spectators. . . .Frankly I wasn't surprised at the good workmanship. It looked great. And it seemed as though the new Hudson was going to create quite a splash. . . .Suddenly all of the frustrations of the last few years were eased. We didn't feel so bad anymore. This was my first experience at milling through crowds who were applauding something I had been a part of, if even in a small way, as a junior stylist. I can say with complete honesty that I was tremendously thrilled."

Andrews later helped develop the Studebaker Avanti under Raymond Loewy, and feels that of all the design programs in which he's been involved, the Avanti's and Hudson's were most similar. It is interesting that both designs are considered among the postwar greats by The Milestone Car Society, which recognizes the 1948-49 Hudson across-the-board. "The similarities in both projects," Andrews says, "were that they were relatively short programs, with the specific direction established early, adhered to closely, with minimum managerial complexities. This gave a strong unity of design without confusion and, immodestly, I think they are both superb automobiles."

Amidst a roar of approval, the new Hudsons for 1948 were eagerly received. "The new car represents the development and refinement of sound design features which have become established through long experience," clichéd A. E. Barit; but even he wasn't prepared for its reception. "A sensational new, ultra-low car—at least 15 years ahead of its time," headlined one newspaper. "The new Hudson hugs the road," said *Popular Science*—people were very big on hugging the road in those days. *Popular Science* continues, "Although the average man can look over the

"This time it's Hudson," was the slogan used as the new Step-down was treated to an overwhelming reception by dealers around the country. Undoubtedly the new car was all the ad flacks claimed, and more.

top, there's lots of room for the big guys in this low-slung car. [There is] greater strength, rigidity, and passenger protection. Going further than Nash and Lincoln, pioneers in the unified body field, Hudson runs the frame girders outside as well as inside the rear wheels, giving, in effect, invisible side bumpers. . . .The car is only five feet high, but [has] plenty of room for even the long-legged and the lowest center of gravity of any U.S. automobile." Said *Time*, "The new Hudson [is] so low that passengers step over the frame and down into it from the curb, yet it still has more headroom and width than any other car now being mass produced."

Wayne Whittaker took one of the first Step-down test drives, for *Popular Mechanics* in 1948. Though his report is a classic example of the press release road test, it jibes closely with the facts. "The first impression is one of sitting closer to the ground than you do in other cars," he wrote.

"Your shoulders are about level with the lower ledge of the window. . . . Now you study the Drive-Master control knob on the instrument panel and are swamped at first by a choice of three shift arrangements: manual, Vacumotive Drive which means manual shifting with automatic clutch, and Drive-Master in which you put the shift lever in high position and step on the gas. . . .The car moves smoothly away. At 25 mph, you release your foot from the accelerator and you are in high.

In December 1948, Hudson leased from the War Assets Administration a portion of the Newcastle, Pennsylvania plant to help ease its steel shortage. The plant produced hand-rolled steel with an output of 5,000 tons per month.

"The car rides smoothly over holes in the pavement that are fully 2½ inches deep. There is no rumble, no rattle and no bouncing around. You are driving 50 miles an hour. You jam on the brakes—the car comes to a smooth and steady halt. There is no pitch. The Hudson maintains an amazingly even keel."

Whittaker quoted Hudson service engineer Milton Bald, who went along with him on the test: "A good many things combine to make a car ride like this. There's the single unit frame and body structure, direct action shock absorbers, the low center of gravity and almost perfect balance. The engine has been moved forward a little and the back seat is no longer over the rear axle but moved forward even ahead of the rear wheels. There are coil springs in front and leaf springs in splayed position—slightly at an angle—in the rear."

Riding in the back seat, Whittaker reported that it gave one a "front seat ride." While driving at 60 mph he passed a trailer truck without a trace of waver. "Suddenly you are shocked to look at the speedometer and see it registering 90 mph. 'Felt like only 70,' you apologize to Bald. Bald smiles: 'A good many people make that mistake.'"

On curves, Whittaker reported "no top-heavy feeling." He accelerated from 10 to 40 mph in nineteen seconds, 10-60 in twenty-nine seconds. Zero to 40, using Drive-Master, came in twelve seconds, and he wasn't able

The 1948 Hudsons in production.

Tevfik Ercan of Bektas, Ercan ve Seriki, was Hudson's Istanbul dealer. Raving about the new Step-down, he is shown demonstrating a '48 in front of Sultan Ahmed I's mosque. This was one of the first Step-downs imported to Turkey.

The 1948 Super Six brougham.

to improve on this time with conventional shift. About the only fault Whittaker mentioned was a tendency to scrape when crossing holes and deep ruts; but despite its five foot height the Step-down Hudson had average ground clearance.

Production was fast increasing by mid-December 1947, Barit announcing that 2,000 units a week were rolling off the Jefferson line. A second assembly line had begun shortly after the new cars began coming off on October 12; six-cylinders slightly trailed eights in initial output. Six months later, Barit leased a government steel mill in New Castle, Pennsylvania: Production, which had risen to 3,000 a week in March, now zoomed to a heady 5,000. If such a rate continued, Hudson would break its all-time record of 300,000 cars in 1929. Initial production consisted of sedans, but broughams were being built by December 10, club coupes by

The 1948 Super Six sedan.

December 17. More that ninety-nine percent of those laid off during the conversion from 1947 to 1948 models had returned to the plant.

Though the convertible brougham was not initially a part of the lineup, it did join the other cars before the model year was over. The full line for 1948 was again divided into Super and Commodore, sixes and eights:

Model Body Style (passengers)	Price	Weight (lb.)
Series 481 Super Six		
sedan, 4dr (6)	$2,222	3,500
brougham (6)	2,172	3,470
club coupe (6)	2,219	3,480
coupe (3)	2,069	3,460
convertible brougham (6)	2,836	3,750
Series 482 Commodore Six		
sedan, 4dr (6)	2,399	3,540
club coupe (6)	2,374	3,550
convertible brougham (6)	3,057	3,780
Series 483 Super Eight		
sedan, 4dr (6)	2,343	3,525
club coupe (6)	2,340	3,495
Series 484 Commodore Eight		
sedan, 4dr (6)	2,514	3,600
club coupe (6)	2,490	3,570
convertible brougham (6)	3,138	3,800

Aside from its unit body, lowered floor and cradled passenger ride, the 1948 Hudson featured many new developments calculated to woo the anxious buyer. Heavy, wrap-around bumpers were recessed into the body and fenders to provide a compact look with no loss in crash protection: the rear bumpers bolted directly to the frame. The windshield, though divided, was well-curved, and offered 758 square inches of area, about thirty percent more than in 1947. The angled side-windows were said to reduce nighttime reflections and aid outward and upward visibility, but the high beltline made that less of an advantage than it might have been.

The instrument panel, though not the prettiest thing ever devised, was nevertheless straight-forward and nicely finished in dark walnut-grain metal, with natural-finish walnut on Commodore center panels that matched the window valences. The radio grille was atop the dash—the first instances of sun-baked speakers were probably encountered on Hudsons—and care was taken to eliminate glare from the red-needled instruments. A two-person rear armrest folded down in Commodore sedans and coupes; broadcloth or Bedford cord upholstery covered coil sprung, airfoam-topped seat cushions. Adjustable front seats tracked upward as they moved forward. Quiet was assured by careful fairing of the skin of the car, reducing wind roar to a minimum by comparison with rivals, and the entire underbody and trunk compartment were sprayed with silica compound to prevent drumming and deaden road noise. There were twelve solid colors offered, and five two-tone combinations.

Mechanical features continued to include the three brake system—hydraulics, mechanical reserve and parking brake—but larger shoes and cylinders were used on the front wheels because of greater front-end weight bias than on 1947 models.

The convertible brougham was announced rather late, on August 19, 1948—assuredly the product of much introspection by Hudson production people. Says Art Kibiger: "Unit construction requires heavy-gauge metal in critical areas, which are usually the ones effected for greenhouse restyling. Convertibles required added frame rails under the car to support the floor when you eliminate the roof structure. The results are great cost and weight increase." Roy Chapin adds, "The only place you could alter [the Step-down body] was above the beltline—without going into extensive retooling." But Stuart Baits says, "There were no problems in building the convertible [on the unit-body] as compared to the frame type. Some additional reinforcements were welded to the underside of the unit body. It was [also] customary to reinforce the conventional frame."

One stylist remembers that Hudson convertibles were literally hand-assembled—because there was no way to make them in the conventional manner: "The convertibles had one room—not a production line—where

The 1948 Super Six coupe.

1948 Super Six convertible brougham.

The 1948 Commodore dashboard, and the Commodore Eight four-door sedan.

they'd be built up. Every one was hand-built. They sawed off the roof of a club coupe—literally sawed it off. On the coupe that roof was welded in the center; this rigidity was lost, so reinforcing [230 pounds more than the club coupe] was added. We had some designs for a four-door convertible brougham during the design stages, though I don't know how many people outside the immediate group were aware of it. But the two-door was as far as we got."

A notable feature of the convertible brougham was its huge windshield header, mainly composed of the leading part of the coupe roof. It was sawed off about nine inches back of the glass, almost obviously to avoid special dies for a convertible header—just as Kaiser did with its sedan-based four-door soft-tops in 1949-51. Hudson advertising made the most of this by claiming safety advantages, though no one ever experimented to see if the

1948 Commodore Eight convertible brougham.

A-pillars that held it up were any stronger than usual. Stuart Baits emphasized the safety angle: "The massive windshield header was primarily a protection for a rollover. . . .Hudson was very conscious of safety and the sedan would survive a roll with little damage and no collapse. The conventional spindly convertible windshield was a disaster."

Hudson convertible production was traditionally miniscule, but whatever the criticisms they should not be considered more than incidental—the cars were visually quite exciting. In 1947, though, only 1,823 soft-tops were built by Hudson; with the late start, the figure was only 1,188 for 1948. There were 3,119 of them for 1949; 3,322 for 1950, but in neither of these years did Hudson count for as much as two percent of total U.S. convertible production. The cars remain rare and—given the particular attention they required in production—valuable examples of hand workmanship in a mass production age.

Commodore Eight convertible tops, available in black or gray, were fitted with Foldaway plate-glass rear lights hinged with vinyl plastic, so that they folded neatly into a leather-paneled and carpeted storage well. (The feature was optional on Supers.) The top's mechanism was hydraulic, so Hudson took advantage of the booster system to provide convertibles with automatic window lifts. Upholstery was Antique red leather, or two kinds of leather and worsted Bedford cord combinations—the cloth specially treated to repel water. The big windshield header housed a Pullman-type interior map-light and radio antenna control knob along with sun visors. It was a comprehensively equipped model, made even more so by Hudson's April introduction of 'valet service'—via a Remington electric shaver which could be plugged into the cigar lighter. Beardless, with the top down, a Hudson driver becomes the living epitome of days gone by.

Production continued to move along briskly, if not as well as management might have preferred, owing to occasional strikes and still-persistent rashes of material shortages. But Hudson dealers had more money than ever—and this encouraged a change in the marketing plan. Hudson dropped its middlemen.

Distributors—in effect wholesalers—had been part of the automobile business for decades. They serviced a network of dealers, bought cars from factories, sold them to retailers and, on occasion, operated as dealers themselves. But as the postwar industry boomed, more and more companies began going to straight factory-dealer relationships. Hudson was the fifth since the war to do so, following Studebaker, Nash, Packard and Chrysler Corporation. Packard did keep six large distributors, including the famed Earle C. Anthony Inc. on the West Coast, mainly for old times' sake. Hudson, possibly with similar motivations, decided to keep a few of its 102 distributors, but the rest were replaced by twenty zone offices supervising ninety percent of Hudson's 2,400 retail outlets.

It was an economical move for the company, of course—a two percent distributor discount would now revert to Hudson, and Jefferson Avenue would have more control over its sales outlets. An irate distributor couldn't defect to, say, Nash, taking a network of dealerships with him. It was an effort to stem the 'raiding' of distributors and dealers by the competition, to better aid local promotion campaigns, and to keep more in touch with the pace of local competition. Dealer defections were beginning to be a factor even this early, despite record sales. As Stuart Baits remarks, "When we had an unusually successful dealer he could not help but wonder about switching, even if he had not been approached."

At least one of the industry's strong personalities feels that the casting out of distributorships was a crucial mistake. James J. Nance, president of Packard from 1952 to 1956, recently told this writer that in Packard's experience it was a short-term benefit with long-term damages. "Where you had distributorships you had pretty good dealers," Nance said, "because they had built that organization over the years." But Packard—and Hudson shortly afterward—had made its decision.

Unfortunately Hudson could not maintain the 5,000 cars-per-week schedule it ran up briefly in May, though the fourth quarter of 1948 saw 51,660 cars, an average of 4,300 per week and the highest quarterly total in many years. Material shortages appeared to be easing, but were still a problem. Sheet steel was in short supply despite Hudson's leased steel plant. Profits, however, were glorious: Hudson made $13,225,923 or $7.28 per share—its best year since 1928. Working capital exceeded $40 million. Shipments for the year, including some 1949 models, totaled 142,454 cars. Calendar year 1948 production was: Super Six 49,390; Super Eight 5,336; Commodore Six 27,160; Commodore Eight 35,314—for a total of 117,200 1948 models. Employment had increased by more that sixty-five percent, Hudson now had over 20,000 people working. And the pace seemed to be continuing: In the usually slow months of January and February, Hudson production was 33,295, almost double the rate of the year before and close to the magic 50,000 car quarter that engendered handsome profits. Barit told the stockholders, "Hudson possesses a superior product, excellent facilities, a strengthened dealer organization and thoroughly experienced factory and field forces. These assets will weigh strongly in our favor in 1949 and cause us to view the future with great confidence."

The 1949 Hudson lineup was as follows:

Model Body Style (passengers)	Price	Weight (lb.)
Series 491 Super Six		
sedan, 4dr (6)	$2,207	3,555
brougham (6)	2,156	3,515
club coupe (6)	2,203	3,480
coupe (3)	2,053	3,485
convertible brougham (6)	2,799	3,750
Series 492 Commodore Six		
sedan, 4dr (6)	2,383	3,625
club coupe (6)	2,359	3,585
convertible brougham (6)	2,952	3,780
Series 493 Super Eight		
sedan, 4dr (6)	2,296	3,565
brougham (6)	2,245	3,545
club coupe (6)	2,292	3,550
Series 494 Commodore Eight		
sedan, 4dr (6)	2,472	3,650
club coupe (6)	2,448	3,600
convertible brougham (6)	3,041	3,800

Why change a good thing? Hudson had it, and Hudson held pat. For 1949, with the exception of a new brougham in the Super Eight series, the company offered its 1948 line all over again. Introduction was in November 1948, and prices were somewhat down over the previous year, reflecting the elimination of initial costs and the easing up of raw materials. But to tell a '49 from a '48, one had to check serial numbers. The most startling new one was a semi-custom by Derham of Rosemont, Pennsylvania, built for Mrs. Roy D. Chapin, Sr.—a division-window formal sedan with leather-covered, padded top and blind rear quarters. Derham eventually built a total of three such formals. The second was on a 1951 Commodore Eight, the third on a 1953 Hornet (Barit's personal car). All three still exist; the first two belong to Hudson-Essex-Terraplane Club members, the third possibly is still in the Barit estate.

The 1949 production body drop. Output soared, but Hudson still couldn't meet its huge demand.

Ground-hog day, 1949 with new Commodore convertible and friends.

Predominately, Hudson advertising stressed the same solid virtues in 1949 as it had in 1948. The Step-down was lower than anybody else's car, yet just as comfortable and with more room inside than any other car. "Hudson Presents," said one persuasive little booklet, "The Importance of Stepping Down"—inside the anxious reader was plied with the true tale of this engineering marvel and the reasons why there simply wasn't anything to compare with it—which there wasn't. But one new factor *had* intervened. In October 1948, the 1949 Nash arrived. Bearing Nash's traditional unit body, it had a startlingly new, streamlined shape. Nash called it the Airflyte. Hudson had other language.

"Don't be confused by Nash's incomplete story on unit body and frame," trumpeted the Hudson *Sales Facts* flyer for 1949. "Only Hudson's exclusive 'Step-down' design (with its recessed floor) and all-steel Monobilt body and frame can give a full measure of all the advantages that are claimed by Nash." While Nash had indeed lowered its car, it had not lowered the floor or, correspondingly, the seats. In terms of roominess, the new Nash offered a healthy 63-inch-wide front seat. But Hudson's two-year-old design offered an inch more, despite an overall width that was an inch less. Hudson's rear seat was a solid foot wider than Nash; there was more headroom, hiproom, legroom and elbowroom in the Step-down, and its driver had about an inch more between the seat cushion and the steering wheel. Hudson had a longer wheelbase—so it rode better—as well as a lower center of gravity—so it handled better. Despite its wheelbase it turned in only twenty feet, against a foot-and-a-half more for its Kenosha rival. "Nash advertising will not confuse you when you fully realize that unit body-and-frame . . .is only part of the story," concluded the Hudson booklet.

It was a good effort, and it made a number of points. There is no question that the 1949 Hudson was a more roadable car and—most would concede—better looking than the 1949 Nash. But it wasn't all one way. Nash's all-coil spring suspension made for a more pillowy ride, a factor becoming salient to buyers of that period. More important, Nash was less expensive model for model, by about $100-200. It was also lighter—though this was not recognized as widely as Nash would have liked in the days of "big cars that hold the road." Hudson, of course, made much of the fact that Nash weighed some 600 pounds less at the curb. But anyone knows now, if they didn't then, that weight does not necessarily improve ride and handling. Nash's surprisingly light unit body bespoke efficient engineering, and many at Hudson realized it. "I can remember when we bought the first Nash Airflyte," says Roy D. Chapin, Jr. "The body weighed half what the Hudson body did. It felt that way—light but rigid. [It was] structurally good, but it felt like a different kind of automobile." Bob Andrews notes, "Hudson was double strength. As I recall the body unit weighed 1,900 pounds. Nash was a little more than half that. Hudson was really over-engineered." Bill Allison, who later designed the Packard torsion-bar suspension, was at Hudson in those years: "The Step-down was so overweight that the first job I had was engineering holes in the structure to try to bring the weight down. This was my only real contribution." One is forced to face the fact that the Hudson was a heavy car. And whatever the weight did for its ride and handling, it was undoubtedly at the cost of money in raw materials—money that was partly reflected in Hudson's relatively high price.

But heaviness did mean solidity and enhanced safety—factors which Hudson had in abundance. And the cars' relatively high horsepower gave them a very favorable pounds-to-horsepower ratio: 28.1:1 for the eight, 29.4 for the six. Nash's lightness gave it a comparable figure of 29.6, but Hudson compared more than favorably with the likes of Pontiac (36.3) Buick Super (33.8), Kaiser (33.5) and the Studebaker Land Cruiser (32.4).

Ed Barit made it clear in mid-1949 that the engines had much more development in them, that the present 6.5:1 compression ratio (7.2 with optional aluminum head) implied no real limit on the ultimate ratio. It could go, he said, as high as fuel would allow, up to and including 12:1.

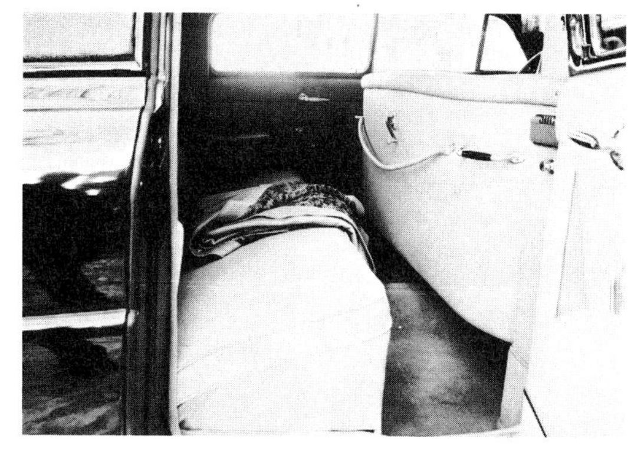

Formal sedan built for Mrs. Roy D. Chapin, Sr., by the Derham Custom Body Company, Rosemont, Pennsylvania.

To illustrate, Barit had a press car increased to 9.3:1 compression and tanked up with 100 octane aviation fuel, allowing the press to drive and evaluate it. Standard automobile fuel octane rating was only eighty-one at the time, and Barit said the oil companies wouldn't raise it more than a point a year. "Does that mean you don't believe 100-octane gas will be generally available for nineteen years?" a reported asked. "Yes," replied Barit. "In that case, what are we doing here now?" "Proving a point, sir."

Hudson celebrated its fortieth anniversary on April 18, 1949, with a nationally-broadcast radio program featuring the ever-popular A. E. Barit, accompanied by sales' George Pratt and Michigan Governor G. Mennen 'Soapy' Williams. Hudson's founders, Barit said, "started out with a total capital of $20,000. A little more than a year ago Hudson reaffirmed its faith in the future of American free enterprise by investing over $16 million to bring out the new Hudson. Today, after producing three-and-one-quarter million automobiles, we view the future with great confidence because of our faith in our ability to grow with America."

For 1949 Hudson duplicated its 1948 performance almost exactly, shipping 144,685 cars. But calendar year production was higher: Super Six 91,334; Super Eight 6,364; Commodore Six 32,714; Commodore Eight 28,688—a total of 159,100 1949's, against only 117,200 1948's. Normally this would indicate growing production and high profits, but net income was only $10,111,219 on net sales of about $260 million. "Most of the reduction in sales volume and earnings in 1949 was caused by the reduced replacement parts and accessories business," Mr. Barit told shareholders. "This is due to the fact that as new cars have become more readily

Hudson wins *Safety Engineering*'s 'America's Safest Car' trophy on its 40th anniversary—it had held the prize since 1941. Left to right are G. H. Pratt, A. E. Barit, publisher Al Best and editor Harry Armand.

available on the market, owners have had less desire to maintain their prewar cars. . . .Another factor is steel procurement. . . .The company was put to even greater expense than [in 1948] to procure steel, [though it] became less of an issue as the year drew to a close and permitted us to discontinue operations of our hand-type steel finishing mill. The need for new automobiles, particularly in this country, remains very great . . .we are in an excellent position . . .we are offering an established car of sound design . . .we are full of confidence."

Nineteen-fifty opened with Hudson out shooting for volume. The first '50 model appeared on November 18, 1949: It was the first major change in the Step-down family, a new, more compact model that Hudson called the Pacemaker. It was hailed, of course, as the dawn of a new automotive vision: "The Pacemaker completes the task Hudson engineers set for themselves four years ago, of making the room, comfort, riding and safety advantages of 'Step-down' design available in a compact car that could be priced within the range of millions of new car buyers." The Pacemaker wasn't compact in today's sense of the word, but it was considerably less bulky than its linemates. Most of the shortening was ahead of the firewall. The wheelbase was 119 inches, against 124 in other Hudsons; overall length was 201½ inches (to 208¼); weight was 3,460 pounds for the sedan against about 100 more for the Super Six and 200 for the Commodore Eight. But the Pacemaker was the same width as the senior cars, giving it an exceptionally broad stance.

The heart of the new car was a 232-cubic-inch version of the Super Six, with a shortened stroke of 3⅞ inches (instead of 4⅜). It produced 112 horsepower at 4000 rpm—same as the top-line Nash Ambassador—and used a 6.7:1 compression ratio with 7.2:1 optional with aluminum cylinder head. Better than half of Hudson's output shifted to Pacemakers for 1950—more than 63,000 of them were built for the calendar year. This was natural, since all Pacemaker models save convertibles listed for less than $2,000. Hudson hadn't been able to claim such low prices since 1947.

Model Body Style (passengers)	Price	Weight (lb.)
Series 500 Pacemaker		
sedan, 4dr (6)	$1,933	3,510
brougham (6)	1,912	3,475
club coupe (6)	1,933	3,460
coupe (3)	1,807	3,445
convertible brougham (6)	2,428	3,655
Series 50A Pacemaker Deluxe		
sedan, 4dr (6)	1,959	3,520
brougham (6)	1,928	3,485
club coupe (6)	1,959	3,470
convertible brougham (6)	2,444	3,665
Series 501 Super Six		
sedan, 4dr (6)	2,105	3,590
brougham (6)	2,068	3,565
club coupe (6)	2,102	3,555
convertible brougham (6)	2,629	3,750
Series 502 Commodore Six		
sedan, 4dr (6)	2,282	3,655
club coupe (6)	2,257	3,640
convertible brougham (6)	2,809	3,840
Series 503 Super Eight		
sedan, 4dr (6)	2,189	3,605
brougham (6)	2,152	3,575
club coupe (6)	2,186	3,560
Series 504 Commodore Eight		
sedan, 4dr (6)	2,366	3,675
club coupe (6)	2,341	3,655
convertible brougham (6)	2,893	3,865

A prosperous Hudson celebrates its 40th anniversary in April 1949. Posing for photographs with a 1909 model 20 and Hudson number 3,250,000, are (left to right): sales vp G. H. Pratt, vp R. W. Jackson, manufacturing vp I. B. Swegles, treasurer A. Hood and purchasing vp G. W. Munger. In the new convertible are Barit (driving), Baits and operations vp H. M. Northrup.

The Pacemaker was trimmed and fitted out frugally. Rocker panel moldings were less ornate than on senior models. The instrument panel was 'simulated fabric-finish,' which neatly eliminated costly woodgrain, and was very plain: utilizing just the bare essentials such as a 'standard' steering wheel without horn ring. Upholstery, though, was wool Bedford cord on seat cushions and backs, the balance done in grey- or Spruce blue-colored Dura-fab vinyl plastic. It was a bid for business in a lower price field. As *Popular Science* noted, the shorter Hudson was $250 less than the Super though still $400 more than the Ford-Chevy-Plymouth range.

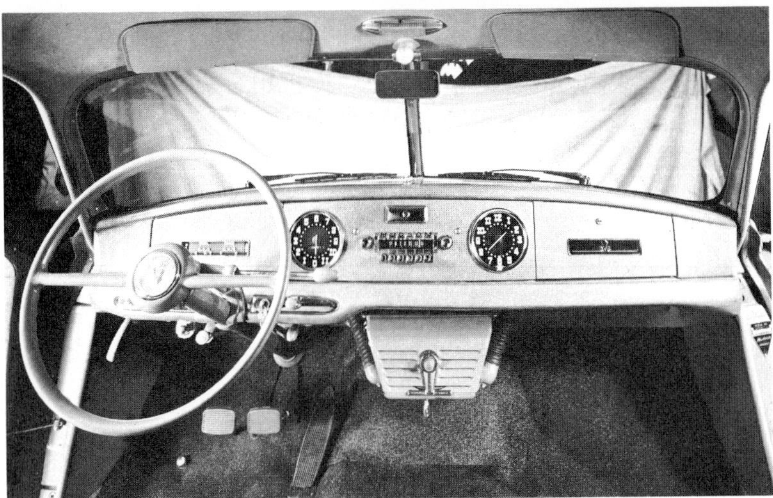

The new Hudson Pacemaker, Hudson's 1950 line leader was on a smaller 119-inch wheelbase and was powered by a 112 hp engine. The business coupe featured a hinged panel behind the rear seat which allowed loading and unloading from inside or outside the car. Storage space totaled 66 cubic feet.

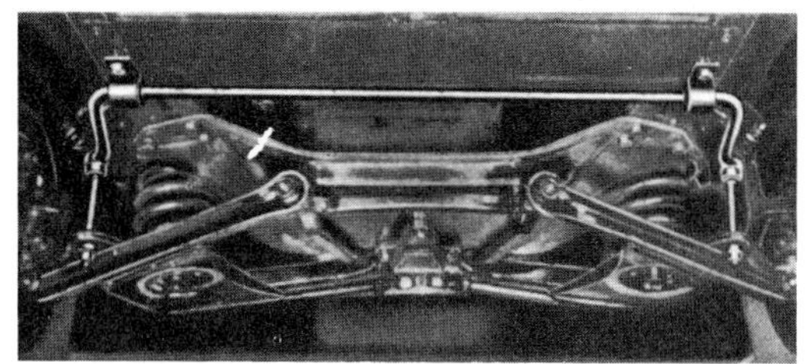

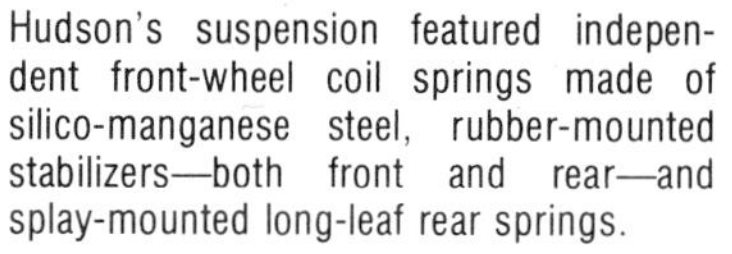

Hudson's suspension featured independent front-wheel coil springs made of silico-manganese steel, rubber-mounted stabilizers—both front and rear—and splay-mounted long-leaf rear springs.

The 1950 Hudsons saw the first Stepdown facelift, with new hood ornaments and diagonal grille braces.

Claimed to be the first of its kind in the industry was a new machine designed to dynamically balance rotating and reciprocating parts of assembled Hudson motors—crank, clutch dampener, flywheel, connecting rods and pistons—while they remained in motion. The Pacemaker, as well as the Super and Commodore, received this treatment. Tappets of all engines were newly-designed mushroom types for even wear, and were pressure-lubricated. The ignition system was improved with a neoprene covered high-tension cable on spark plug wires, and was water-, heat- and oil-proof.

The senior models, introduced February 10, 1950, were again very slightly altered. The Super Six lost a three-passenger coupe, but otherwise the model lineup was identical to 1949. Grilles were revised—four horizontal bars instead of five were used, with diagonal struts forming the Hudson triangle motif. Parking lights on Supers and Commodores were faired into the bottom bars; grillework was brought forward slightly and there were new hood and fender ornaments. On Commodores, the fuselage crease was further emphasized with a stainless steel molding underneath it, extending from the rear fenders to just above the front wheel housing, where it ended with a Commodore nameplate.

Three-dimensional patterned nylon Bedford cord was a new Commodore feature—tan with brown stripes or blue-grey with blue stripes. The senior cars were offered in eleven solid and four two-tone color combinations.

Overdrive was a $99 option on all 1950 models, but the Vacumotive Drive automatic clutch had been eliminated. Drive-Master remained an option at $105, but most Hudson publicity this year went to a new $199 transmission accessory called Supermatic. This differed from Drive-Master

The 1950 Commodore Eight sedan.

1950 Commodore Eight convertible brougham.

by offering a high cruising overdrive ratio. Like Drive-Master, one started by placing the shift lever in normal high position and lifting the foot when he wished to shift into regular drive. The high cruising overdrive gear was automatically brought into play (with the dashboard Supermatic button engaged) at a speed of 22 mph. For reserve power on hills or during passing Supermatic automatically downshifted to a 'reserve power gear,' and on decelerating to 10-13 mph it shifted back into first. An emergency low gear was also provided for heavy-duty driving and hard pulls.

"Hudson Supermatic Drive is economical," said Hudson, "giving up to three miles free out of every ten the car travels. It gives fast acceleration since action is immediate and positive. The driver has complete control of the car at all times. With Supermatic Drive there is no tendency for the car to creep or move forward at the stop light. Therefore, it is not necessary to apply the brakes when waiting for the traffic light to turn green. Due to definite clutch engagement, towing a trailer or another car is practicable."

Nineteen-fifty was, by most industry yardsticks, a very good year. There was no indication of the usual January sales downturn, and price cuts on such new models as Buick, Cadillac, Oldsmobile, Studebaker and Willys as well as Hudson (which cut Commodores and Supers by up to $166.50 in late spring) only inspired more sales. When a strike hit Chrysler in the first quarter, competition swarmed in to take up the slack. Some 2,400 Hudson employees were added in April; retail deliveries that month were ten percent ahead of 1949. The firm sold a record 15,927 cars in April, against an output of only 11,516; 18,023 were sold in May to 11,868 produced—so inventories were rapidly declining. West Coast dealers were flying customers to Detroit to drive their new Hudsons home, applying the normal shipping charges to a transcontinental vacation. There still seemed no end to the unbelievable seller's market, now in its fifth year.

Hudson work stoppages were slight in 1950—six-and-one-half days due to material shortages, five days worth of wildcat strikes, a week for the 1951 model changeover. By November, the first anniversary of the Pacemaker, 69,000 of these models (1950's and 1951's) had left the Jefferson Avenue plant.

April 1950 saw the resumption of car production at Hudson's Tilbury, Ontario plant—the first since before the war. Hudson Motors of Canada Ltd., which had been founded in 1932, had sold cars on an import basis since 1946; local production was expected to increase Canadian volume.

All this activity nonetheless did not see any significant improvement in the Hudson position. Calendar year production, at 143,006, was almost

exactly what it had been in 1948 and 1949, but in an improving market both Chrysler and Kaiser-Frazer moved ahead of Hudson, dropping it from eleventh place in the production race to thirteenth. The Pacemaker, intended to draw some sales from low-middle competitors like Mercury and Pontiac, was significantly eating into Super Six sales. At the same time, eight-cylinder Hudsons were on the wane. Thus, 1950 model year production was considerably less than 1949:

Model	1949	1950
Pacemaker	——	63,252
Super Six	91,334	17,246
Super Eight	6,364	1,074
Commodore Six	32,714	24,605
Commodore Eight	28,688	16,731
Total eights	35,052	17,805
Total sixes	124,048	105,103
GRAND TOTAL	159,100	122,908

A 1950 Super Eight as most frequently viewed by other drivers.

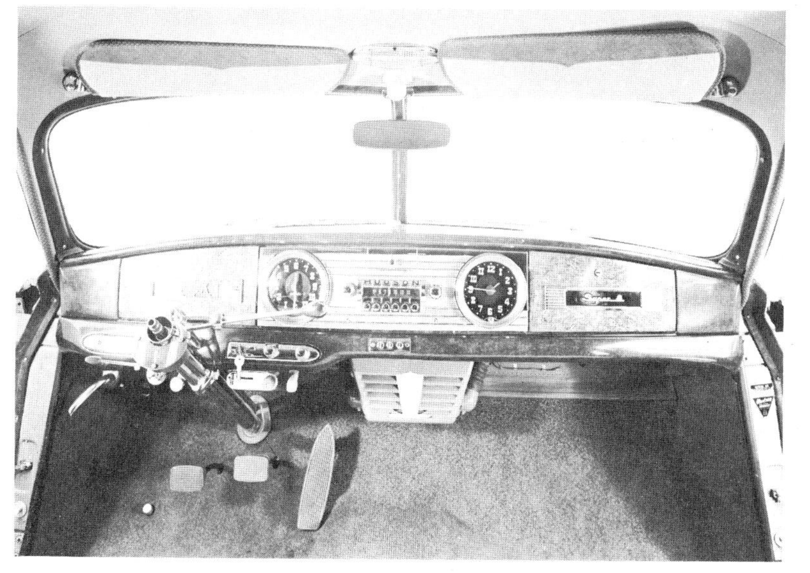

The Commodore Six and Super Eight dashboards for 1950.

Hudson's plants in Detroit, Britain and Canada, 1950.

Calendar year output remaining approximately the same, Hudson could record another profit of above $10 million—the figure exactly was $12,002,274 on sales of $267 million—a slightly improved profit/sales ratio compared to 1949. Shipments during 1950 had amounted to 143,586. Because of the Korean emergency, the company found itself negotiating defense contracts in early 1951, though Barit promised "as many automobiles as we are permitted" under the circumstances. And there was real worry that the industry might be halted again, as in 1942.

At least one former Hudson executive holds the 1950 performance critical. "It should have been plain after 1950 that things just couldn't go on the same old way forever. The Pacemaker came in fully $400 below our topmost model and was in brand new sales territory. The plant had a capability of 300,000 a year without much strain. Why didn't production leap up in 1950? My guess is that the market for the larger Step-downs was becoming saturated, the sellers' market in general was not as strong by then

A Step-down being loaded for export.

A fully equipped 1949 Commodore Custom Eight set a new Daytona class record of 98.63 mph in August 1949. T. Marion Smith set the record after race officials had called off a motorcycle event because of rough beach conditions. Here Smith receives his award from a race official.

as we'd thought. The Pacemaker was a stunning success—were it not, we would have been comfortably in the red. So instead of forging ahead we were breaking even. At this point the inability of the now three-year-old unit body to undergo radical change began to make itself felt. By 1951, it would begin to cost us money."

Hindsight is cheap and far too easily indulged. Whether the same individual felt that way in 1950 is impossible to tell, and even he doesn't remember. It was a good year financially—Hudson hadn't seen a $10 million plus profit since 1929, now it had run up three such years in a row. There were other hopes, other distractions. In August, for example, a fully-equipped 1949 Commodore Eight driven by Daytona dealer T. Marion Smith set a new Daytona Beach speed record for its class: 98.63 mph. The record was set after officials had canceled a motorcycle race because the beach was too rough—not too rough, apparently, for the Commodore. The car was timed by the new National Association of Stock Car Auto Racing—NASCAR—and around NASCAR circuits word began to spread of the big, powerful, utterly stable cars being turned out by Ed Barit and his friends on Jefferson Avenue.

That record-breaking run at Daytona would prove to be prophetic.

CHAPTER 3

The Fabulous Hudson Hornet

ONE NIGHT AT Hudson, PR man Frank Hedge was working late: "I remember it distinctly because we had lady policemen to guard the place. One of them came up to the office. 'There's a man here named Marshall Teague,' she said. 'He's been trying all day to get into Engineering. He's waited and waited, but no luck. And he has to see somebody.' I'd never met Marshall Teague so I didn't know him from Adam. He wanted some help with some manifolds. He was entering racing, he said, and there had been little pieces about him in the paper.

"To make a long story short, I took Marshall and a friend of his to dinner that night. The next day I phoned Engineering and got them to help get Marshall some manifolds. Hudson's advertising director, Bob Roberts, was scratching around for some excuse for publicity then, and this appealed to him. Suddenly Advertising was really interested, and I was traveling around the country to these races. And then we started winning. . . ."

Past doubt. And Hudson kept winning. For four straight years, 1951, 1952, 1953 and 1954, the Hudson Hornet and its new H-145 six was the *force majeur* in American stock car racing. A late start gave Hudson only third place in the 1951 NASCAR standings, with eleven wins against Oldsmobile's twenty, but for 1952 and 1953 it was virtually unbeatable: Hudsons finished first in twenty-seven out of thirty-four NASCAR Grand Nationals those two years. Even in 1954, with new, V-8-powered rivals swarming into the lists this six-cylinder car (ranking eighth in industry horsepower) was first in eight out of sixteen Grand Nationals and finished in the top five positions no less than forty-three times.

"Racing, in a way, probably was one of Hudson's saviors," says Roy D. Chapin, Jr. "At least [it] tended to minimize the V-8 question, because the performance image was there even with a six, which of course was a helluvan engine." As the British say, Roy Chapin is spot-on. In a period of rapid transition to the overhead valve V-8 in performance cars, the Hudson Hornet was an anomaly. The sole 'big six' in the country, and the last, it was not even an overhead valve design. Some say it was the biggest six we've ever built—not so, Packard built one of 525 cubic inches back in 1912—but it was certainly in a class by itself in the early fifties. If it bespoke a certain conservatism on the part of Hudson management and engineering, it nevertheless flew in the face of convention and all probability, by truly trimming such 'progressive' iron as the Oldsmobile 88 and Chrysler Saratoga on the huge ovals of NASCAR and the AAA. Said the latter's contest board news bureau director Russ Catlin, when announcing Hudson's hold on every AAA national and track record in 1952: "The fact that Hudson dominates our activity is only incidental—but I think certainly worthy of heartiest congratulations."

This writer asked Stuart G. Baits why Hudson decided to bore and stroke its way to horsepower with a six in 1951, rather than switch to a new V-8 like almost everyone else. "Hudson had six-cylinder manufacturing

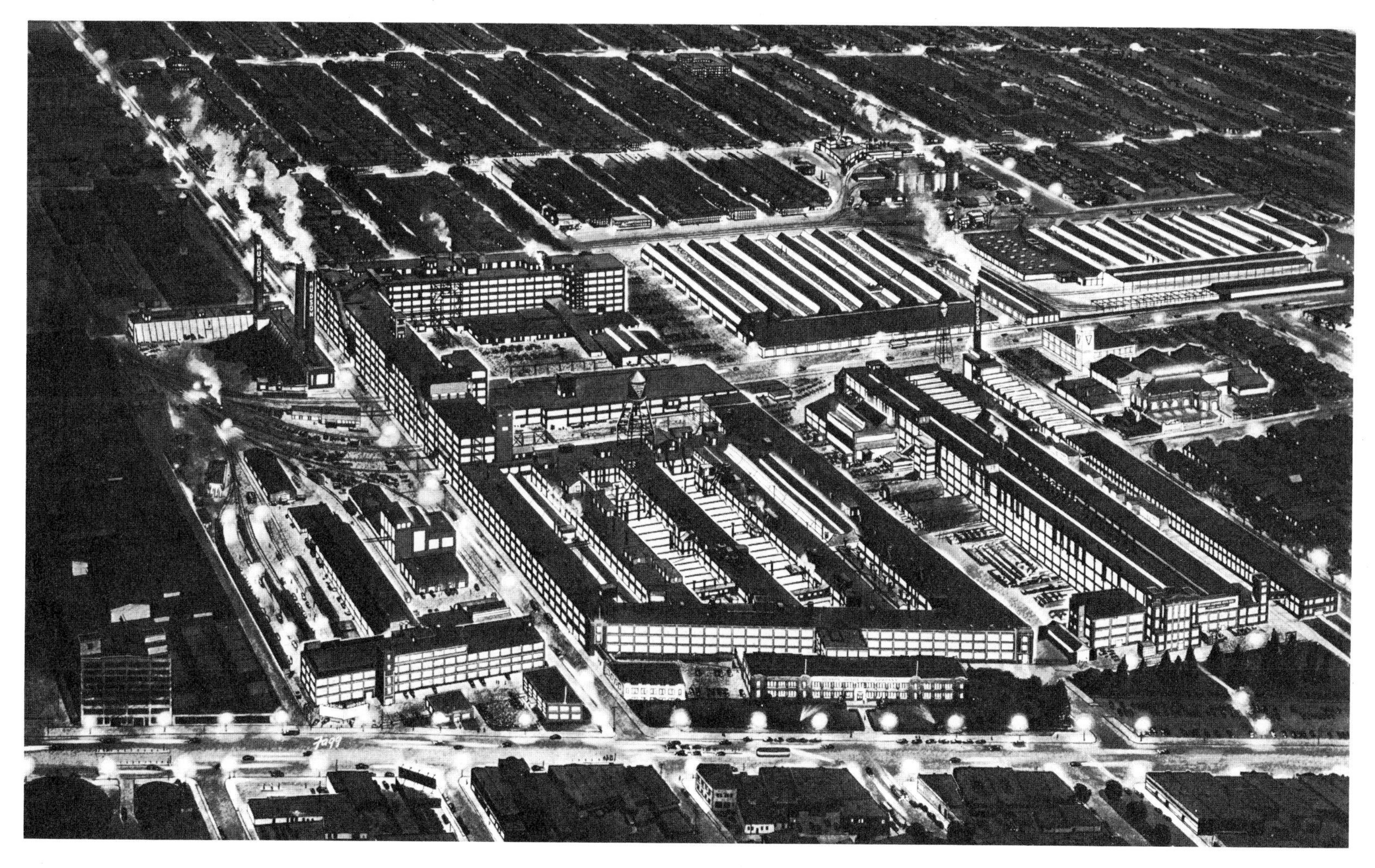

"One night at Hudson, Frank Hedge was working late." Striking nighttime view of the East Jefferson Avenue plant at the height of Hudson's prosperity.

equipment," Baits explained. "The V-8 was not yet so dominant at the time and cost was certainly a factor. This motor was very powerful, and it dominated stock car racing against V-8's, overhead valves and whatever else the competition was using. Please also note that the six-cylinder engine, recently pronounced dead, is very much alive now that efficiency is important."

In an earlier interview with the Hudson-Essex-Terraplane Club's Guy Folger, Baits gave credit to "Dana Badertscher, who was our engine designer on the Hornet, for producing the outstanding design."

The 1951 Super Custom six sedan.

The casual often assume that the H-145 engine was a sort of *deus ex machina* from Jefferson Avenue. It was nothing of the sort, of course. In its understressed, easy-revving form and its granitic durability, it was not unlike Hudson's Super and Special Sixes of thirty or forty years before. And, as Maurice Hendry notes, "This reversion to type carried a penalty. Superior performance was undeniable, but whether the eight's level of refinement had been maintained, let alone surpassed, was something else. Although the engines were all statically balanced before assembly, then dynamically balanced electronically, there were complaints from owners and road testers that the new engine was neither as quiet nor as smooth running as the old—hardly surprising in view of the increase in displacement and [the two less] cylinders."

Plainly, the H-145 was a moderately bored and stroked evolution of the previous 262-cubic-inch Super Six which, by the way, was continued in the 1951 Super and Commodore Sixes:

	Super Six	**H-145**
bore	$3\frac{9}{16}$ in.	$3\frac{13}{16}$ in.
stroke	$4\frac{3}{8}$ in.	$4\frac{1}{2}$ in.
displacement	262 cu in.	308 cu in.
horsepower	123 @ 4000 rpm	145 @ 3800 rpm
torque	198 lb/ft @ 1600 rpm	275 lb/ft @ 1800 rpm

The last specification was obviously H-145's most important distinction—thirty percent more torque was offered than on the Super Six. There was also eighteen percent more horsepower, a 10 mph increase in top speed, a second or two less time to accomplish the standing start quarter-mile. With its standard 7.2:1 compression aluminum head, the H-145 could propel a factory stock Hornet from rest to sixty in thirteen seconds, and exceed the 100 mph mark. Hudson's continued adherence to 'hard' chrome alloy blocks reduced the need for frequent valve grinds and assured long motor life by reducing valve, tappet and bore wear. Oil consumption by Hudson engines remained satisfactory for much longer, on average, than engines with 'softer' cylinder blocks.

The new engine reduced Hudson's pounds/horsepower ratio from an already good twenty-eight or twenty-nine to just 24.8. Combined with the immensely strong Step-down unit body, the result had to guarantee impressive performance. It did. In 1951 it was simply unmatched by any American car, and few Europeans except sports-racing machines. Men like Marsh Teague soon recognized its potential and came calling—to a pleasant reception as it turned out—at the Hudson Motor Car Company.

The Hornet effectively replaced the Super Eight, leaving the Commodore as Hudson's only eight-cylinder model in 1951, while the Pacemaker line shrank to a single series, priced about $200 more model-for-model than the 1950 Pacemaker Deluxe. The 1951 lineup:

Model Body Style (passengers)	Price	Weight (lb)
Series 4A Pacemaker Custom		
sedan, 4dr (6)	$2,145	3,460
brougham (6)	2,102	2,430
club coupe (6)	2,145	3,410
coupe (3)	1,964	3,380
convertible brougham (6)	2,642	3,600
Series 5A Super Custom		
sedan, 4dr (6)	2,287	3,565
brougham (6)	2,238	3,535
club coupe	2,287	3,525
Hollywood hardtop (6)	2,605	3,590
convertible brougham (6)	2,827	3,720
Series 6A Commodore Custom-Six		
sedan, 4dr (6)	2,480	3,600
club coupe (6)	2,455	3,585
Hollywood hardtop (6)	2,780	3,640
convertible brougham (6)	3,011	3,785
Series 7A Hornet		
sedan, 4dr (6)	2,568	3,600
club coupe (6)	2,543	3,580
Hollywood hardtop (6)	2,869	3,630
convertible brougham (6)	3,099	3,780
Series 8A Commodore Custom-Eight		
sedan, 4dr (6)	2,568	3,620
club coupe (6)	2,543	3,600
Hollywood hardtop (6)	2,869	3,650
convertible brougham (6)	3,099	3,800

Nineteen fifty-one's new body style, offered in four flavors via four different model lines, was the Hollywood hardtop—built like the convertibles, rather individually, and beefed-up to absorb the stresses of a pillarless window line. The Hollywood must be reckoned a response, rather than an innovation. Chrysler had introduced the first hardtops, albeit sparingly, on its 1946 Town & Country; GM followed on the 1949 Oldsmobile, Buick and Cadillac and on Pontiac/Chevrolet in 1950. Kaiser-Frazer developed a four-door hardtop with permanent glass window-pillars in 1949. Ford, Chrysler, Studebaker and Packard, as well as Hudson, all rushed onto the hardtop bandwagon in 1951.

The 1951 Pacemaker sedan.

On the Hollywood, Sales Vice President N. K. VanDerZee's pitch reached crescendo proportions: "The Hollywood provides the ultimate beauty in automotive design. We believe that no other car can match its graceful lowness of design that combines the dash and verve of a convertible with the luxury and convenience of a custom-crafted Club Coupe." The car was accentuated, he added, "by the liberal use of chrome." He wasn't fooling. There was a new chrome beltline, windshield pillars, windshield divider, header and rear window pillars. A special two-tone color option of Jefferson green and Corinthian cream was offered, along with all the regular and optional Hudson colors: thirteen solids and four two-tones.

Though the Hollywood was a good looking job on an already very sleek, low-built car, the 1951 Hudson was basically the same old suit with

'51 Commodore Custom six Hollywood hardtop.

Commodore Custom six convertible, 1951.

1951 Commodore Custom eight sedan.

a new set of buttons. The grilles were more massive, three-bar affairs; the top two bars bowed to meet the bottom with parking lights built into the apex, and slim central diagonal struts framed the illuminated Hudson emblem. Hornets and Commodores used double rub-rails, one at the body side, another atop the rocker panels. A special Hornet emblem, resembling the insect not so much as a rocket ship, was mounted on the quarter panels and deck. Frank Hedge believes the Hornet was named by the ad department. American Motors later revived it for a new line of compacts—and 'Wasp,' which Hudson introduced in 1952, was in the running for what became the Pacer until AMC realized what the acronym had come to stand for by 1975.

On the seats, Hornets and Commodores featured a specially designed three-dimensional nylon insert panel. Instruments were regrouped on a more curvilinear dashboard. Mechanically the H-145 engine was major news, but Hudson also acknowledged a growing customer perference for fully automatic transmissions by contracting with GM for their four-speed Hydra-Matic. This became a $158 option on all models save Pacemakers, which retained Supermatic availability. Supermatic was dropped on the higher-priced cars, though Drive-Master remained a $98 extra, and overdrive $100. Hudson's patented Triple-safe braking system was retained.

Marshall Teague, who probably did more than anyone for Hudson's great racing image, poses with his 'Fabulous Hudson Hornet' in 1951.

Motor Trend's Walt Woron was among the first testers of the Hornet. Taking its nameplate literally, Walt announced that "It hums, buzzes, zooms and [does] practically everything else any self-respecting hornet would do, except fly. [But] the top speed of this car isn't far from approaching 'take-off' speeds. The Hudson Hornet is impatient—it wants to go, and runs best fast. It's powerful, fast in the getaway, is easy to handle and gives a comfortable ride." Woron coaxed his Hydra-Matic-equipped Hornet from nought-to-threescore in fourteen seconds, ran the standing quarter-mile in 19.4 at 70 mph using low and high ranges, and topped out at close to 98 mph.

The speedy 1951 Hornet sedan and a view of its hunkered down rear quarters.

Though the Hornet was a large car, Woron felt it was "quite easy to maneuver and to judge distances (even though you cannot see the right front fender from the driver's position). There's plenty of leg- and headroom, and all operating controls are handy. Two disconcerting features . . .are the speedometer location and the rearview mirror. The speedometer is on the far left of the instrument panel and is hard to see with your left hand on the upper section of the steering wheel. The rear window makes cars appear squatter than they actually are—and by raising or lowering yourself in the seat, you can cause them to change their shape."

Woron was enthusiastic about unit construction. "If you've ever seen a Hudson after it has been involved in an accident, you've seen the greatest advantage to the unit type, all-welded frame and body. . . . The safety aspect of this arrangement cannot be denied, while the merits of the 'step-down' floor are dependent on personal likes and human configurations."

Woron's Hornet averaged about eighteen miles per gallon at an economical 30 mph; Floyd Clymer, who produced one of his famous *Popular Mechanics* surveys on the car, reported figures of up to 21 mpg from owners, though their average was about the same. Clymer, never much of a Hudson backer in the past, admitted that his test Hornet had definitely changed his mind: "The Hornet is a wildcat on wheels, due to its very high horsepower-to-weight ratio. It is a dynamic performer which responds to the throttle like a thoroughbred. . . .The low center of weight . . . makes it a fine road car and it is noticeably steady at all road speeds. There is no road wandering, it takes the corners more like a racing car than a stock job and handles over rough roads with much ease of operation." Clymer also approved of the twin-carburetor option on the Hornet (in 1951, this meant two Pacemaker carbs on a dual-intake manifold) and special head such as one with 8:1 compression on the Super models. "A Pacemaker head installed on a Hornet will give a ratio of engine compression as powerful as 9:1." The owners he talked to echoed his praise: A high seventy-four percent said they'd definitely buy another Hudson, and only one percent definitely would not. Objections noted mainly dealt with limited trunk space, poor interior and exterior finish and some bad body fitting, but seventeen percent of all owners found no objections at all. Step-down construction, visibility, interior finish and styling all won over ninety percent approval. A solid ninety-four percent of the owners rated their car from good to excellent. Yet another tribute came from the American Society of Engineers in February 1951, when it presented its Merit Award for "the most advanced ideas in design" to Hudson for the second year running.

Unfortunately the Hornet, Hydra-Matic and Hollywood stood Hudson insufficiently in a generally declining year for the industry, though the Hornet did account for 32.5 percent of sales. The company thus recorded its first loss since 1939: $1,125,210 on net sales of only $186 million. Ed Barit explained that production was hampered by a shortage of sheet steel at mill prices, delays by the government's Office of Price Stabilization in the approval of price increases to cover rising costs, and strikes "which

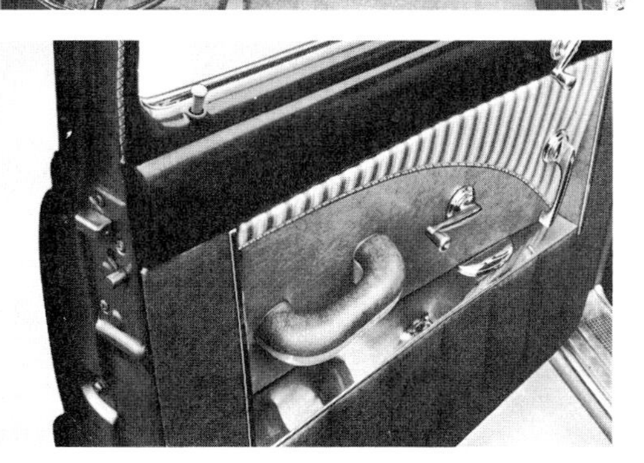

The 1951 Hornet Hollywood hardtop.

The 1951 Hornet convertible and interior views of the 1951 Hornet.

The Hudson assembly line, final production area, 1951.

occurred intermittently in our plants during the greater part of the summer." Anxious to improve efficiency, Hudson had embarked on time studies and recommended new work procedures, which caused a whacking labor-management stand-off. Workers would report on time, but on the assembly line (subject to the new work changes) they would soon start skipping operations. Finally management would stop the line in disgust and send the men home. Hudson lost several weeks' production until Federal mediators settled the disagreements in August, and even then, a tenuous truce prevailed. Production suffered, more so than at the big manufacturers. Chevrolet and Ford, for example, recorded decreases of about twenty-five percent in their output, and Plymouth actually gained a little, while Hudson's drop to 93,327 units for calendar 1951 represented a decrease of thirty-five percent and a drop to under two percent of the new car market. Of the 133,697 1951 model year Hudsons, 43,666 were Hornets, 36,277 Pacemakers, 22,532 Supers, 16,979 Commodore Sixes and 14,243 Commodore Eights.

The government further hampered car sales in an already (because of the Korean conflict) uncertain year by restricting financing agreements for new car purchases to eighteen months. This action by the Federal Reserve Board was protested by thousands of car dealers, who charged that it was destroying the ability of middle income families to buy cars. (One Colonel Blimp in the FRB stated that "the nation must have a depression at once.") The independents insisted that the Board was "abusing credit powers to support monopolistic tendencies . . .with the Big Three destined to do 90 percent of the business."

The credit crunch gradually eased, though it had done irreparable damage, and toward the end of 1951 the Office of Price Stabilization finally approved Hudson's requested higher prices for the 1952 models. Its cash position increased from $10 million in December to $25 million in March 1952, and during 1952's first quarter the company returned to profitable status. Barit predicted that 1952 would be "a year of realization." In many ways he was right. It wasn't known then, but Hudson was never again to record a 100,000 car year.

Meanwhile, the company could console itself with a major bright spot for 1951: the Hornet, and its track performance. It all stemmed from the efforts of that friendly, heavy-set Floridian, Marshall Teague, and the cooperation of enthusiastic Hudson engineers.

Teague ran a service station and garage in Daytona, where he tuned engines to a fine pitch and closely followed neighboring NASCAR racing. His relationship with Hudson established, he convinced North Carolina truck farmer Herb Thomas to switch from the Oldsmobile 88's he'd been driving (with mixed success) to Hudson Hornets. Gregarious and unassuming, the two worked closely with Hudson engineering and even made personal appearances at Hudson sales drives. "Teague was a fine driver and mechanic," said Stuart Baits, "very modest as are all the good drivers I ever knew, easy to work with. And jointly we produced cars that dominated stock car racing. . . ."

Marshall Teague insisted that without any cheating, through hairline timing and racing clearances, he could get 108 to 112 mph from a Hornet certifiable as strictly stock—and he did it in 1951, before the announcement of many Hudson 'severe usage' components or Twin H-Power. What's more, Teague allowed a public account of how he did it in the May 1952 issue of *Motor Trend,* in a story by Eugene Jaderquist.

"Everything came apart," Jaderquist reported, "—engine, transmission, differential, wheels, steering mechanism. [One] underwent a change of transmission as well—Marsh prefers to drive standard gears rather than Hydra-Matic . . .The other Hudson kept its Hydra-Matic because Herb Thomas feels perfectly at home with modern transmissions.

Herb Thomas and Marshall Teague with Hudson's Tom Rhoades at Daytona.

Thomas chasing Fonty Flock's Oldsmobile at Jacksonville, 1951.

"Most important part of the job is to 'free everything up' as Marsh puts it. NASCAR specs allow a .033-inch overbore so the maximum was used on this car and a minimum of .002 inches was left between pistons and cylinder walls. [Piston rings] were replaced by Grants, which will seat in 20 minutes. Wrist pins were freed in the pin holes; rod and main bearings were adjusted and set to run without using excessive oil but so they were completely free. This last operation takes about eight hours of painstaking checking and rechecking. Final clearances are set with the aid of a plastic compound." One thing Teague didn't divulge was his method of valve timing, though he admitted that the job took more time than most tuners would spend. "The stock engine is *approximately* correct," Jaderquist reported. "Marsh makes it *exactly* right.

"Before buttoning down the head, the capacity of each combustion chamber must be measured. . . .If one chamber is larger or smaller than the others, the compression ratio of that cylinder will be different, the amount of push on the power stroke will be different, and the engine will be out of balance. A rough-running plant, fighting against itself, doesn't win.

"The 'freeing-up' process is now applied to wheel bearings and the entire drive train. Axles, transmission and rear-end gears are thoroughly checked for flaws because one of the primary causes of stock-car breakdowns is failure of these parts. Coil springs from the 'export' kit are installed with heavy-duty shock absorbers, from the same kit, inside. Steering tie-rods, all beefed-up to withstand racing stresses, are thoughtfully provided by Hudson, but Teague goes even further and has the new parts shot-peened. This process hardens and strengthens the surface of the metal."

After qualifying, last minute preparations would be made. "The rear seat was removed, which is mandatory, and the rear cushion was taken out also. Headlights were taped, the hood was anchored firmly by stout, elastic cord, and the doors strapped shut. On the dash panel, Marshall's brother marked 49 chalk lines to represent laps. As Marsh approached the south turn each lap he would rub out one line out with his finger—a simple, effective way of keeping track. Hubcaps were removed. The muffler had already been removed for the qualifying run and it remained off for the race."

Herb Thomas nips the verge—shown at Daytona in 1951.

Knocked out of orbit by Herb Thomas's thundering Step-down, Hershel Buchanan's Nash spins at Columbia, SC, 1952.

Thomas and Hudson at Langhorne, PA.

In this day of scientific 'prodifying' to the tune of many thousands of dollars, such modifications appear almost 'backyard' by comparison. Even in 1951, they were much less than most NASCAR competitors insisted on. As such, the Hornet's racing record can only be considered a phenomenal tribute to the car itself—as it left the factory. "Nothing was done to [Teague's car] that you couldn't do with your own Hornet," Jaderquist concluded. "What it amounts to is simply cleaning up some of the factory's rough work, adjusting to hairline accuracy a few of Hudson's approximations. If the factory did it for you, your Hornet would cost at least $500 more, not counting the price of the special 'export' kit. Hudson, and all the other Detroit firms believe you'd rather save money than race. After all, very few of us will ever drive a Hudson to Speed Week."

"Everywhere I go people ask me why the Hornet wins so consistently," said Marsh Teague. "There are really three important reasons—durability,. safety and power.

"With a victory at Phoenix, Arizona [April 22, 1951], I moved into the point lead for stock-car race drivers. I felt then that each race is the equal of 50,000 miles of hard highway driving. So you can see a Hudson can really take it. My Hornet stands up under brutal punishment.

"My faith in Hudson's stability was proved beyond any doubt at the Gardena, California stock car race [April 8, 1951]. Although the Gardena track was only a half mile job, my Hudson Hornet took me to the front of the pack in the first lap, and I gained distance on every turn. The car almost drove itself; it glided through the turns as easy as you please, all because of

A young Roy Chapin, Jr., in suit, congratulates Tim Flock after Flock's eighth 1951 victory (left). Hudson driver Dick Rathmann with Bob Calvin (right).

Hudson's lower center of gravity. Hudson has a big safety edge when it comes to brakes. . . .The finest hydraulics are backed by a reserve mechanical system.

"Hudson power is the talk of any track. The acceleration pushes you out in front of the pack in a hurry. . . .And once out front, the H-145 has everything it takes to keep you there. . . .Power must be smooth as silk, too. The hour-in, hour-out pounding we give a car makes any misbalance or vibration stand out like a sore thumb. The H-145 is . . .perfectly balanced."

Teague's co-driver Herb Thomas had his first Hudson ride at the Detroit 250 in August 1951. He lasted just twenty-six laps and finished fifty-seventh out of sixty cars. But it was a false start. Three weeks later Thomas romped home an easy first at the significant Southern 500 at Darlington, South Carolina. The Teague/Thomas duo was established.

Herb Thomas was equally enthusiastic about his Hornet. "A Hudson was in the lead 356 of the 400 laps," he said. "I grabbed the number one spot on the ninety-fifth lap and poured on the power to stay in front all the way in . . .at the finish I was six laps ahead of the third place car. Jesse Taylor, also driving a Hudson Hornet, finished second.

"It was a scorcher of a day; I thought the black asphalt track was going to melt. A lot of cars did wither on the vine, but my Hornet seemed to thrive on punishment. [My time] beat all records by more than eight minutes. And the surprising thing is, I didn't have my Hudson wide open at any time during the race. But that extra power was always there to get me out of a tight spot. That's a safety feature a lot of people don't appreciate.

"I gained a lot of distance in the turns. The car hugs the road like it loves it and has riding qualities and stability I've found in no other car. The engine . . .gives smooth power no matter what speed range you drive in. And it's got tremendous acceleration. I still marvel at how it stood up to the tortuous grind. Over 6½ hours of constant punishment, competing against everything else the industry could offer and Hudson took first *and second.*"

The lanky Carolinian's innate modesty forbade him from commenting on the significance of his personal achievement—every bit as important as

Enshrined at the Joe Weatherly Museum, Darlington Speedway, South Carolina, is Herb Thomas's 'Old 92' Hornet. It started 41 Grand Nationals, won seven and carried Thomas to his first championship.

that of his car. He vanquished some of the fastest people on the ovals—Johnny Mantz the 1950 winner, Lee Petty, Tommy Thompson, Tim and Fonty Flock, and Curtis Turner among a field of seventy-five starters—and established a new speed record of 76.57 mph for the 500 miles. After Darlington, Thomas vied with Fonty Flock for the NASCAR title in the remaining races. Ultimately he was declared champion, after Grand National wins of 100-150 miles each at Langhorne, Pennsylvania; Charlotte and Occoneechee, North Carolina; Macon, Georgia; and Jacksonville, Florida. Fonty, also Hudson-mounted, was second with an equal number of victories, 143 NASCAR points behind, while brother Tim was third—in another Hornet.

In November big Marshall was back in the headlines, with a fine performance in the second Carrera Panamericana—that car-breaking 2,000-mile open road race over the worst imaginable Mexican roads from the Guatemalan border to Texas. Though overwhelmed in the straights by Ferraris, Teague incredibly held them on the corners, to the *caramba!* of spectators and amazement of Alberto Ascari and Piero Taruffi Teague's Hudson finished a respectable sixth, less than an hour behind the Italians.

Roy Chapin, Jr., was one of a group of Hudson people waiting to greet Teague at the finish, as he roared up in a cloud of dust and applause. "How I remember that race," Chapin says. "One of the other drivers was saying, 'Marshall, I was watching you and you were turning 6000!' Marshall asked what he meant and he said, 'I just stood up in my car and looked through at your tach. That thing really turns up, doesn't it?' Can you imagine going 120 mph down a Mexican road close enough together to read the instrument panel of the car in front of you? I remember Teague getting out of that car. The first thing he said was, 'Where's the bullhring?' He'd been sitting behind that wheel five days straight and all he wanted to do was go see a bullfight! Those were some guys. . . ."

The 1952 season came, and before it had passed into history "the Fabulous Hudson Hornet," as the admen called it, had become the dominant stock car. When it wasn't first, it could usually be assumed that the drivers had plastered one or two of them against a wall in the attempt—and other Hornets were usually second or third. Rivals insisted the cars were far from stock—the Lincolns which began competing in the Carrera in 1953 were so highly modified they were mechanically unrecognizable from production—but the 'Teaguemobiles' were almost innocently prodified. They were aided by special racing options, but these were nowhere as abundant as those that Lincoln had. The Hornet's real forte was great driver-mechanics and a cooperative Jefferson Avenue. Of the Hudson Company Teague said, "They know more about the car than any of us. If you want to get statistics on how many cars of how many makes have started in a race, their finish positions and failures, don't ask the race steward, ask the factory. They have all the data and it doesn't come from newspaper reports."

Best-known Hudson 'severe usage' item was Twin H-Power, formally introduced as an option in 1953 but a device Hudson had had on the shelves since 1944—and the firm had experimented with dual carburetors since 1937. Twin H consisted of a dual manifold, dual carburetor induction system, providing a very even measure of fuel-air mixture to the cylinders, which in a standard induction arrangement is not ideally suited to six-cylinder configurations. It was a fair reply to the four-barrel carburetor then coming into use on competitive V-8's.

Hudson claimed that Twin H-Power was the first instance of dual carb, dual manifold induction on an American-made six: "The secret of the amazing results which Twin H-Power provides is that it so accurately measures the gasoline, so evenly distributes it to each cylinder, and so thoroughly vaporizes the fuel with air that it provides what the engineers call far better 'breathing' and combustion than has heretofore been obtainable.

"Most important, this greatly improved efficiency is obtained on regular grade gasoline. There is no need to pay for premium grade fuels. (Twin H-Power) provides jet-like pickup and a sustained flow of terrific energy . . .gives you stepped-up power where you need it most, in the ordinary driving range.

"The extreme ruggedness of the Hudson engine is an important factor in the development of Twin H-Power. The oversize bearings, the extreme rigidity and hardness of the block, the weight and stiffness of the crank-shaft—in fact the extra sturdiness built into all parts—make it possible to utilize the extra power"

Other competition-oriented equipment was announced in Hudson Parts Merchandising bulletins as follows:

May 21, 1952: twin carburetor kits for Wasp and Commodore Sixes at $125 retail.

November 15, 1952: severe usage parts kit (rear axle, shocks etc.) for Wasp, Commodore Six and Hornet, $211.12.

January 2, 1953: severe usage rear axle and hub/bolt assemblies for Pacemaker, Wasp and Hornet, also usable on 1952 six-cylinder Hudsons, $182.50.

January 5, 1953: special cylinder heads for any model "to take care of an Owner's requirements," $34.35 to $51.25. Also twin exhaust manifold and pipe kit for all models, $50.

January 8, 1953: skeleton engine assembly for all Hornets "which provides extra high performance for use in high altitudes, police work, etc. This assembly includes cylinder head and gasket. It uses the manifolds, water pump, oil pan, flywheel, carburetion equipment, electrical units, etc. from the engine being replaced." Retail price $385.

April 17, 1953: severe usage radiator assembly for Hornet.

June 8, 1953: extra high-performance cam for Hornet, $32.50

August 27, 1953: extra capacity fuel tank.

April 13, 1954: new twin-carburetor kits for new models, $125 to $133.

As late as the mid-fifties, the skeleton 308 high-performance engine, Twin H-Power kit with oil-bath air cleaners, special exhaust system and camshafts were still available from American Motors. This is not to mention a variety of non-Hudson-built performance parts that proliferated in the Hornet's wake, and to a large extent are still available—for Hudson is still winning races in the most unlikely circumstances.

In 1952, the Hornet was invincible. Tim Flock led off with a win at the Grand National opener at West Palm Beach on January 20. Teague then won the Daytona Beach Grand National driving a new two-door brougham, as usual absolutely stock. He started thirteenth, took the lead on lap four, and was never headed, driving without a pit stop and averaging 74.65 mph. He was followed by Herb Thomas, and sixty other cars. Of only thirty-seven ultimate finishers, six were Hudsons.

All-Hudson front row and start of the 1952 Southern 500.

In March, Marshall Teague won and Hornets finished 1-2-4-5 at Jacksonville (an Olds 88 was third). Herb Thomas won a Grand National in March. Then in April at Martinsville, Virginia, Dick Rathmann made it five wins in a row for Hudson; and just to prove it was no freak victory,

April 1952: Dick Rathmann takes the checker at the Martinsville, Virginia, Grand National—fifth successive Hornet win in the year's first five races.

Rathmann came back to win again at Langhorne on May 3. The Pennsylvania circuit was the fastest mile-long dirt track in the country and provided a rough challenge, but Rathmann was never out of first place from the second lap, and by the finish he was six *miles* ahead of Tim Flock's runner-up Hornet, which in turn was five miles ahead of the other eighteen finishers.

By May, Hudson held first place laurels in ten out of eleven Grand Nationals—along with five seconds, two thirds, two fourths and a fifth. Six months later Hudson had won forty-four races, including several victories by Marsh Teague in AAA events, which he had begun entering in May. Teague won the AAA championship in 1952 with seven victories, finishing with a margin of more than 1,000 championship points over his nearest rival. In overall AAA competition Hudsons were winners twelve times in thirteen events.

More Hudsons—288—competed in stock car races during 1952 than any other make. Plymouth had 185; Olds, 174; Ford, 132. No other makes saw battle more than thirty times. Despite its high numbers, Hudson had a respectable sixty percent finishers—only Plymouth among its major rivals was significantly better, with seventy-eight percent finishers. But Plymouth wasn't even close in the standings. With twenty-seven NASCAR wins it was Hudson's year, hands down.

At a Detroit meeting in September, Hudson parts merchandising manager H. L. Templin spoke of "the great value that Hudson has obtained from stock car racing. The publicity covering Hudson's many wins has been of untold benefit to us." Templin stressed the need for Zones to be familiar with severe usage parts and for dealers to keep them in stock. "Zone stocks of racing parts are not complete, due to the unfamiliarity of the parts and part numbers of this special equipment."

While Hudson admittedly made the most of its racing success in advertising, one marvels at its almost casual approach to the events themselves. Not only were many dealers completely unfamiliar with the special components, the factory itself was never very evident at trackside—as so many factories are today, with dozens of vans packed with parts, and scurrying little men dressed in white coats, carrying clipboards. How different Hudson's attitude was, even then—and how successful the cars were nonetheless—is a source of constant wonder to this writer and, doubtless, to anyone who ever watched the big Hornets compete.

Diminishing slightly, the wins continued in 1953 and 1954, thanks to Templin's special parts and the highest development of Hudson racing engines, the 7-X. A dealer-installed option, the 7-X engine provided the ultimate in Hudson performance during these last two years of the fabulous Hornet.

Mr. Pete Booz of the Hudson-Essex-Terraplane Club had the luck to discover a low mileage, genuine 7-X engine in 1965, and has since researched the story of this modification in detail. He provides the following lucid description of how it differed from standard: "The blocks of all 7-X engines were overbored a minimun of .020 inches, with two larger overbores available, and were deeply relieved. A high-lift cam was installed, but it burned up timing chains and was too noisy; it was often replaced with cam 311040, the last cam designed for a Hornet engine, according to factory documents. The 7-X also used a special hand-reworked head from the 232 engine. The 232 head gave a higher compression, and the reworking done to it was to take best advantage of the larger 7-X valves.

"The 7-X also featured a studded block to hold higher compression afforded by the special head, Twin H and an oddball split-dual exhaust

Marshall Teague was the national champion for 1952, finishing the AAA schedule with a 1,000-plus point margin over his nearest rival, and capturing 12 of 13 AAA stock car events.

manifold and header pipe. Special work was done on the block around the valves, to take advantage of their larger size. One member of the H-E-T Club who feels he is an authority on the subject rates these 7-X factory engines at 210 hp.

"It is important to note that the 7-X was a dealer-installed option—could not be ordered in a car from the factory according to dealer bulletins. Also, most of the few that were sold came as short block replacements, rather than as a totally new engine package. There are very few original 7-X engines around. Most of them are Hornet engines people have tried to modify to 7-X specs. They can be spotted, even if they have the usual 7-X exterior hardware, by checking the serial number stamped on the right front corner of the block. If, as in all factory-installed engines, it matches the chassis number, it is not a genuine 7-X. The 7-X blocks had to be hand-stamped by the dealer. For instance, the few here in California all have 'CAL' followed by some numbers so they could be properly registered. This was done by the dealers when they sold the engines."

The 7-X 'skeleton' engine assembly was officially announced on August 3, 1953, though it had seen competition before that. H. L. Templin introduced it with the traditional claim that it had been engineered "for police cars and high altitude work." While some highway patrols did indeed use Hornets, this was a standard rejoinder to qualify such hairy options as 'stock' items. At $385 the package included cylinder head assembly, crankshaft, valves, knurled pistons, camshaft, connecting rods, timing chain, cam and crank sprockets, valve springs, tappets and adjusting screws.

There is no question that the 7-X was an outstanding engine development—probably the highest state of tune for an L-head six in automotive history, and a credit to Hudson's excellent engineers. But the company seemed too satisfied with it. It was one thing to keep developing the big six, yet quite another to let it forever compete for sales with rival V-8's. The Chrysler hemi, for example, was selling extremely well, and showing tremendous performance potential by 1954 in the hands of modifiers like Briggs Cunningham. It was only a matter of time before its superior developmental potential would tell. Still, though one former Hudson employee claims that a proprietary V-8 was contemplated, Stuart G. Baits firmly insists that "no V-8 engines were considered."

Counting every race, Hudson scored twenty-nine victories in 1953, but its NASCAR total was down to twenty-two. In 1954, with twenty-one wins overall, Hudson won only seventeen Grand Nationals. The Hornet remained NASCAR champion both years, but the decline was as obvious and ominous as the rising horsepower of the Chrysler hemi. Motor

Others in Hudson's racing hall of fame: (left to right) Herb Thomas, Frank Mundy, Al Keller and Dick Rathmann.

pressmen quickly pounced on the theory that the Hornet had essentially 'lost its sting.' Developed to the limit and all strung out, they said, it was no match for V-8's in similar stages of development.

Yet the Hudson Hornet continued to do well, and still does. With the AAA's abandonment of racing in 1955, many marks set by the Hornet remain today. Furthermore, its victories in certain recent amateur contests have caused more than a few raised eyebrows. Consider the experience of Bruce Mori-Kubo, an enthusiastic Minnesota owner/driver . . . "One of our most satisfying wins was in 1971 at the Basso Presto Sports Car club Hillclimb in St. Peter, Minnesota. We won the sedan class with a time of 29.41 seconds, beating the first place Class D Triumph Spitfire which scored 33.26, the first place Class C Datsun which had 31.17, the second place Class B MGB with 29.48, and the sixth place Class A Corvette with 29.48. Another happy event was the Gene Conners Memorial Hillclimb in 1969. We won this time in Class D with 33.90, ahead of the Class C Porsche (35.51), the Class B Pontiac Firebird, and several Porsches and MGB's plus a few Class A cars."

If this is all so unbelievable, take heart. It was pretty unbelievable in the mid-fifties, too. Unquestionably, the Fabulous Hudson Hornet was the most hairy-chested, impressively performing six the American industry has ever created. It stood athwart automotive history, against many very good engines of V configuration, and firmly said 'no.' The best epitaph it could have are the hundreds of examples still running, still racing today, and the words of Don Vorderman: "Remember, the next time you encounter one of these harmless looking hump-backed mastodons with 'Twin H-Power' on the trunk lid, you're witnessing the golden years of a retired champ, with all the integrity and commanding every bit of the respect that is due to a Joe Louis or a Ted Williams."

Anyone who feels that this is overstatement is respectfully requested to consider the appendix listing of Hudson's racing records.

Something which many Hudson fans regard quite as important as the Hudson racing record was the cars' sheer invulnerability to the ravages of time. In *Consumer's Report's* 1955 analysis of "frequency of repair," for example, Hudson topped the list with the least amount of work done to the cars by their owners—and the only car that was even close to Hudson's record was Ford. Such eminent reliability caused many, many more Hudsons to survive into the sixties and seventies, to the benefit of enthusiasts and collectors of a later era.

Back in Detroit, most companies made little more than serial number changes between 1951 and 1952, and Hudson followed the general trend. Its alterations were the mildest since the 1948's turned into 1949's: The hood ornament was moved forward, rub rails extended and swept downward, taillights lowered and wrapped around, the rear window enlarged. Hudson introduced optional Solex tinted glass, which reduced ultra-violet rays by fifty percent; offered key-start ignition; mounted the spare tire upright to get more trunk space; and replaced the 124-inch-wheelbase Commodore Six with a new model called the Wasp, using the Pacemaker's 119-inch wheelbase. The Wasp ran an H-127 six of 262 cubic inches—the same engine that had been offered on the 1951 Commodore, though its horsepower was up slightly. Pacemakers were available in eleven colors, other Hudsons in eleven solids and fifteen two-tones. The 1952 selection:

Model Body Style (passengers)	Price	Weight (lb.)
Series 4B Pacemaker		
sedan, 4dr (6)	$2,311	3,390
brougham (6)	2,264	3,355
club coupe (6)	2,311	3,335
coupe (3)	2,116	3,305
Series 5B Wasp		
sedan, 4dr (6)	2,466	3,485
brougham (6)	2,413	3,470
club coupe (6)	2,466	3,435
Hollywood hardtop (6)	2,812	3,525
convertible brougham (6)	3,048	3,635
Series 6B Commodore Six		
sedan, 4dr (6)	2,674	3,595
club coupe (6)	2,647	3,550
Hollywood hardtop (6)	3,000	3,625
convertible brougham (6)	3,247	3,750
Series 7B Hornet		
sedan, 4dr (6)	2,769	3,600
club coupe (6)	2,742	3,550
Hollywood hardtop (6)	3,095	3,630
convertible brougham (6)	3,342	3,750
Series 8B Commodore Eight		
sedan, 4dr (6)	2,769	3,630
club coupe (6)	2,742	3,580
Hollywood hardtop (6)	3,095	3,660
convertible brougham (6)	3,342	3,770

For the enthusiast of the fifties, there was no better way to watch stock car races than from the roof of your Hudson.

The inimitable Tom McCahill's remarks on the 1952 Hornet will perhaps be of interest. Our beloved Unk didn't equivocate: "Hudsons are America's finest cars from the very important standpoint of roadability, cornering and steering. . . .A car like the Hudson, that has fairly fast steering and cornering ability, will keep its feet under it and be a lot safer, even at 20 mph, than a typical current Detroit balloon that just mushes and loses its footing under a fast wheel cut. [Hudsons] are able to duck. The balloons, like the statue of Sheridan's horse, usually have to stand up and take it."

The '52 Hornet Tom drew for his *Mechanix Illustrated* road test was more than interesting—a Teaguemobile that had benefited from the magic fingers of Big Marsh himself. The venue, too, was singular: the circuit at Le Mans, France, site of the world's longest endurance run where the Hudson served as Briggs Cunningham's 'staff car.' (An interesting fact in itself, since Cunningham's racing engines were of GM and Chrysler manufacture.) Cunningham's troops, McCahill said, had "literally beat the brains out of this Hornet until it was as loose as the morals around the Place Pigalle. I am giving you this background in fairness to explain why the car went rear-endless before my tests were over."

Could the Hornet, king of American oval tracks, handle one of the world's toughest road courses? Silly question. "I racked up about 200 miles of high-speed driving around Le Mans during informal practice sessions before the big race," Uncle Tom reported "I found that I was getting down the [Mulsanne] straight at around 107 mph, which is fast, but the real pay-off came in the turns. The Hornet stayed right in there with some of the practicing race cars on the corners and was actually faster than some of them in getting out! I took the famous S-bend at White House corner at open throttle and had no trouble in staying on the tail of a Lancia Aurelia that was to race a few days later.

"This is America's best road car by far, in spite of its Queen Mary width which I for one could do without. As I reported several years ago, the windshield is so far forward from the driver's eyes that you get the feeling you are always coming out a tunnel. This and a few silly blotches of chrome are all the bad points I could dig up. Actually the Hudson is one of the finest automobiles built in America, regardless of price. It will match any American car in comfort, roominess and general appointments. If you like 'em big and tough, here's your dish."

The 1952 Hudson Wasp Hollywood hardtop and 1952 Wasp convertible.

Here are McCahill's performance figures with Cunningham's Teague-tuned Hornet, along with his own stopwatch checks on a regular model, and similar figures from 1952 road tests by *Road & Track* and *Motor Trend* (Asterisk indicates Hydra-Matic):

	McCahill Teague-tuned	McCahill standard*	Road & Track*	Motor Trend
0-60 (sec)	12.1	14.7	14.8	16.8
top speed (mph)	107.0	96.5	97.8	99.2

Road & Track, never wont to praise Detroit iron, with a corresponding tendency to get all dewy-eyed about a fifty-seven horsepower M.G., awarded the Hornet high marks. "The Hudson is not my dream car by any means," said John Bond, "but if your wife wants a bigger car than the Joneses' and you want performance, durability and better than average roadability, the Hudson Hornet is the car for you." Oliver Billingsley, even further out in the sports car orbit, praised Hudson's styling: "The body contours of the entire Hudson line were developed by Hudson stylist Frank Springer [sic; sorry Frank, sorry Art] and consultant Reid Railton to give both streamlining and high strength. Altho the Hudson will never be considered a 'classic' design . . .too much chrome and stuff . . .it is one of the cleanest designs on the market. I particularly like the absence of 'rocket-jet-spaceship' influence and the conservative rear fender treatment."

Motor Trend also tested a 1952 Wasp, feeling that it and the Hornet, on paper, were the two most interesting of Hudson's five models. Walt Woron found the Wasp, with lower horsepower and weight, only 1.4 mph slower than the Hornet: "At the top range of speed each of the cars has a certain amount of 'floating' sensation, with the Wasp almost giving you the sensation of being airborne. At anything up to top speed, however, both cars were very stable." The Wasp, Woron said, could be thrown into turns with the same abandon as the Hornet. "This car rides considerably softer than the Hornet—it has more float, more bounce and more sway. The suspension is, undoubtedly, set up softer to compensate for the shorter wheelbase. Road rumble was very evident." *Motor Trend* was the only tester to do well on mileage, however. They achieved 16.6 mpg with the Wasp at a steady 60 mph and 22.7 mpg at a steady 30. With the Hornet, *Motor Trend* reported 16.2 and 21.3 respectively. *Popular Science*'s Wilbur Shaw could squeeze only 14 mpg out of the Hornet, *Road & Track* averaged only 15 mpg—but decided Hornet buyers wouldn't worry about high fuel consumption anyway.

This Wasp, in this writer's opinion, is an underrated Hudson—especially given today's driving conditions. So many Step-downs are still in moderate or daily use, one should venture an opinon as to their practicality for those to whom Detroit doesn't speak anymore. The Wasp is very practical. Its compact wheelbase and better fuel consumption are attractive features. Its ride, if more pillowy than the Hornet's is still far, far more stable than the typical 1952 Detroit car, and it can be flung into corners with little fear. It comes in all the usual Hudson models including the Hollywood hardtop and convertible brougham. By the time it arrived Hudson styling had settled into perhaps the cleanest form it ever took on the Step-down. And it's cheaper to buy today than a Hornet, though the latter is by no means high-priced.

He really shouldn't say this, but as a non-owner all his life, the author could now readily subscribe to the White Triangle faith—a very peculiar feeling that he doesn't usually have. It's something an automotive writer should never do—rather like a doctor falling in love with a patient—but reading about, riding in, driving and writing of Step-down Hudsons in the preparation of this tome has made a believer out of a lifetime advocate of little cars, and Packards.

While 1952 was not a good year for car production, Hudson's 76,348 units for the twelve months was enough to put it back ahead of Kaiser-Frazer (71,306), and into fourteenth place in the industry. In corporate production Hudson was sixth, behind the Big Three, Studebaker and Nash, but it was comfortably ahead of Kaiser-Frazer, Packard, Willys and Crosley. Its industry share was almost exactly the 1.75 percent it had been in 1951.

By the end of the year, over 4,000 people were working on defense contracts at Jefferson and other plants—mainly for Wright R-3350 aircraft engine components, B-57 rear fuselage and tail assemblies and RB-47B forward sections. Of Hudson's $214 million in sales, $24 million was garnered from this sector. The magic 50,000 car quarter was a long way off in 1952—the best three months was January-March with 22,500 units. Nearly one-third of production volume toward the end of the year was in Wasps. Government control of strategic raw materials had been lifted after the initial Korean emergency, but sheet steel remained in short supply, as Hudson put it "extremely difficult and expensive to procure. Hudson, along with other large users, finds it necessary to resort to unusual measures to secure a sufficient supply. A number of strikes are now in effect in suppliers' plants and are threatening production in the immediate future."

A. E. Barit, whose duty it was to remain optimistic, remained optimistic. "Our engineers and our consultants, both here and abroad, are engaged in many forward projects aimed at the continued development of our cars and maintaining their competititve superiority. One of the projects involves a new type of small car . . .I am sure that when this car is exhibited, which will be done shortly, it will prove of interest to the public."

That was a surprise, coming from a company whose product line didn't even include a station wagon. But by the time of Barit's report, Hudson already had a product to hang such a title on: what he called "a new kind of car. [It] has characteristics of compactness which make it especially adaptable to present day road and parking conditions. Also because of these characteristics it is capable of unusual performance and has splendid roadability. Because it sells in the lowest price field it greatly broadens our coverage of the market, enabling our dealers to compete for a very large percentage of automobile purchases. Additional body types are either in the tooling stage or projected for the future and will be announced when ready for the market. This new car will be increasingly helpful to our business as time goes on."

That the new car would be anything *but* helpful to Hudson's business wasn't known then. A compact, on a short, 105-inch wheelbase, was just what the public wanted. Hadn't the public already proven this, by its reception of the Nash Rambler and the Henry J, which together sold close to 200,000 copies in 1950-51? Of course! And Hudson's small car was stronger, faster, better than the Rambler or the Henry J. Besides that, it had an excellent name, one to which everybody in America could relate, though Oliver Billingsley probably didn't like it.

Hudson called it the Jet.

Hudson's Edsel: The Short, Sad Flight of the Jet

THERE ARE FEW generalities in the auto business, few rules by which one can solidly predict a product's success or failure. In an ideal world, car sales would be directly proportionate to engineering and styling quality. But this is not an ideal world, and though these *are* factors, the single most unpredictable determinant is the marketplace. The marketplace is composed of people. Auto companies are continually spending millions to try to understand what people want. Those who say Detroit has been 'dictating' our tastes were not keeping the books at Ford when the Edsel arrived, or at GM during the Corvair's presence, or at Chrysler during the early fifties, when relatively shorter, more compact cars almost caused that company's demise—or at Hudson during the short tenure of the Jet.

As a product, the Jet is closely comparable to the late lamented Edsel. Both were soundly designed and had many good engineering features. Both were very quick. Neither was very pretty, and their looks didn't stand either car well in the sales race. But most important, both Edsel and Jet were ill-suited to the times—the former a big, flashy car in the borning years of the small car revolution, the latter almost the diametric opposite in an era of big cars. Both were colossal management mistakes. In Ford's case the mistake cost a hundred million dollars. In Hudson's, it cost the company its life.

"I think the thing that really caused the downfall, if you want to call it that, or accelerated it at least, was the Jet," says Roy D. Chapin, Jr. "This should have been a successful small automobile. The dealers were just crying for a car of that type. At that time I think I was a regional manager when the Jet was coming into being—1952 or so. We were promoting it with a drawing—that was all we had. In the drawing (car illustrations in those days were 'stretched,' and the people were scaled-down to make the cars appear larger) it wasn't a bad-looking car. Unfortunately the finished product didn't look like the drawing. It was high and narrow and bulged in the wrong places. It's a shame, because conceptually the car was excellent. I think by the time they got through making all the compromises, many of which I'm sure were necessary, it lost practically everything it started out to be and wound up a product relatively costly to build in relation to the price you probably should get for it." The writer reminded Mr. Chapin that the basic Jet came in at $1,858 stripped—against $1,670 for the cheapest Chevrolet, $1,690 for Ford and $1,750 for Plymouth. "That," he replied, "was a very real problem."

How did Hudson make such a serious mistake? Well, its management hadn't really distinguished itself in the immediate past, had it? Stylists have related how the Step-down was, in the end, something less than they wanted it to be. But the Step-down was beautifully engineered (some would say over-engineered) and a tremendous performer, and these factors saved it somehow. Regardless, almost any sort of car could be sold in the great seller's market. But by the advent of the Jet, the senior Step-downs hadn't

Ed Barit with the first production Jet, January 2, 1953.

The Jet had a heavy, rugged unit body like that of the large Hudsons.

Jet engines also offered a Twin H-Power induction system optionally.

changed very much for six model years, and in performance, roadability and looks the competition was starting to get a shade rough.

Indeed, Hudson management seems never to have had an eye for style, despite the best efforts of Frank Spring. Hudson engineering insisted on constructing cars as if they were part of the foundation of the Brooklyn Bridge. This might have been all right in the upper-medium price field where the Hornet was situated, but in the compact car region it was a mistake. No matter how good the Jet was on the road—and the evidence is that it was very good—it simply cost too much to build. Its styling was poor by comparison to the sleek Aero-Willys and even the Rambler. That it didn't sell could be traced in part to these factors.

The marketplace was, however, the main factor. Had the times been different—on the order, say, of 1959—the Jet might have sold like nickel hot dogs. The reader may recall Studebaker's Lark, another homely little car, cheaply built and best known for its ability to rust. Introduced for 1959, the Lark turned Studebaker's fortunes around and saw the company to a profit for the first time in five years.

Most sources agree that A. E. Barit's go-ahead for what became the Hudson Jet was largely predicated on dealer demand and a careless look at the success of the Rambler and Henry J—careless because those two pioneer compacts didn't really do as well as their initial sales figures had promised. The Henry J sold over 80,000 units for 1951—but it was introduced in February 1950, so that particular model year was eighteen months long—and it sold only 32,000 copies for model year 1952. The Nash Rambler's score was better: 20,782 in 1950; 57,555 in 1951; 53,055 in 1952. The Rambler was the only small car selling at that level *and* selling fringe models at that—convertibles, wagons, hardtops—but there was no real indication of a great underlying demand for others like it. The best-selling imported economy car in 1952 was the British Austin—with a grand total of 4,490. There was a lot of talk about compacts, mainly among the motor magazines, and one supposes that Hudson dealers competing heavily with other independent franchises like Nash and Kaiser-Frazer probably caught a little flak for not having a Rambler-like car. But mostly the great compact market which drew the attention and channeled off the funds of Nash, Kaiser-Frazer and Hudson was much overblown in those

Steyr version of the Fiat 1900—similar to the 1400 model which probably inspired Ed Barit in the styling of the Jet.

The Jet as Frank Spring conceived it, from a drawing by Terry Boyce.

days. The only successful one of the three was Nash, with the Rambler, and even that model remained a minority of Nash output until 1955, when the small car market was truly born. Any market survey worth its salt would have disclosed that for Hudson even to match Rambler's production with a car like the Jet would require the automotive equivalent of raining up. But Hudson apparently never undertook a market survey. . . .

So Barit's first mistake was building the Jet at all. Given the state of the market, its timing would have been enough to kill it, but that was accompanied by a second mistake—Jet styling.

Mike Lamm, of *Special-Interest Autos*, researched Jet design in 1971 while preparing a feature on the Italia sports car, unearthing a surprising uniformity of opinion. Clara Spring said of her husband: "He was very disappointed when [the Jet] went through the works. It came out like a little box on wheels, and it wasn't smart and it wasn't at all low and compact." A Spring associate, Victor Racz, said, "The Jet was [Spring's] biggest beef. He had originally designed it four inches [some sources say three inches] lower, and then the sales force said, 'Well, we want the car higher . . .we don't want to go way out.' So it came out looking like a 1952 Ford."

A contemporary Fiat also influenced the design of the Jet. Says former Hudson engineer J. J. Hartmeyer, "Mr. Barit liked the seating arrangement of the Fiat, so that's why he insisted that the interior package be [that way], and that's why the damn thing was so high. Spring's idea was to drop the height, to get it down lower like the Mercedes. But Mr. Barit wanted those chair-high seats. And he used the Fiat as the package guide."

The Fiat in question was probably the 1400. Introduced in 1950 and current through 1958, it logically would have been around just when Barit began thinking small car. Its dimensions are extremely close to those of the production Jet:

	Fiat 1400	**Hudson Jet**
wheelbase (in.)	104.3	105.0
track, front (in.)	52.2	54.0
track, rear (in.)	52.2	52.0

The 1400 body was of unit construction, which could only have appealed to Hudson people—and 120,000 examples were built, which must have appealed to them too. But the 1400 was powered by a small, four-cylinder engine and was proportionately lower and lighter than the Jet—and easily better-looking. The Jet's exterior was staid: a high beltline, confusing body sculpture, a narrow stance, almost an inch higher than the big Hudson. For 1953 this was poor testimonial to Step-down design; Studebaker, without a Step-down body/frame, managed that year to build a car fully seven inches shorter than the Jet.

One outstanding dimensional plus on the Jet was its interior space utilization. The high build couldn't help but create ample headroom, and with 58-inch-wide seats against a narrow 67 1/16-inch overall width, the Jet utilized an astounding eighty-six percent of its overall width for passengers. The back seat, however, was cramped. All of these factors were present in the Fiat as well and, as Hartmeyer continues, "The seating package started with the Fiat. Our first test car was built around the Fiat's proportions. We even put a Fiat nameplate on the thing so if somebody walked up and saw the name, he wouldn't start wondering what the hell kind of car this was."

The 1953 Super Jet sedan.

The Ford influence was apparently provided via Chicago dealer Jim Moran who, according to Roy Chapin, was peddling "about five percent of Hudson output—he was selling 3,000 Hudsons a year and it was an unbelievable thing. He used to come out and sit in the styling discussions, and complain about lack of change [on the big cars]." Moran espoused some features of the 1952 Ford. Says Hartmeyer, "The Jet originally had a small rear window in it. Ford came out with a wrap-around window, and Mr. Barit and this Chicago dealer had us change to a wrap-around. Pittsburgh Plate Glass said they could wrap it if we held the Ford curvature. Mr. Barit wanted high fenders—they were lower when the original Jet was first styled. Mr. Barit like the round taillights of the Oldsmobile." The rear quarters of the car evolved into high, slab-sided affairs, capped with little hemispherical Olds 88-type taillights.

Some reports have the Jet a close copy of the 1952 Ford, which Hudson enthusiasts rebut with the improbability of lead times: As a 1953 model, the Jet would have had to have been locked-up in styling before the 1952 Ford appeared. But if Hartmeyer is correct in stating that PPG could wrap the Jet's rear window if it followed Ford's contour, it is not unlikely that Hudson knew about the Dearborn car prior to its introduction. The '52 Ford appeared in January of that year—plenty of time for a die change to achieve the wrap-around rear window, a relatively minor alteration. The lack of a rain channel at the juncture of window and roof indicates some hasty last minute alterations. It is important to note that Hartmeyer and others do not say the Jet was a copy of the Ford, *except* for its rear window. The most they say was that it *resembled* the Ford, which is quite another matter.

Unfortunately it is hard to tie down the Jet's styling evolution because Hudson perversely threw away or destroyed all evidence of its prototypes. The few photos and models that do survive are items Hudson overlooked. *Special-Interest Autos* did construct a composite of what Spring's original Jet looked like, basing the effort on comments from those interviewed. It is probably quite accurate. Notably, it uses a fuselage and roofline not unlike the big Hudsons. Where it differs from the production Jet—and this has been substantiated by several sources—was in its much lower line to hood and deck. "If you look at a Jet today," Mike Lamm wrote, "you'll see a horizontal crease in the hood about two-thirds of the way down its nose. This crease was Frank Spring's original hood line. Likewise, around back there's another crease in the decklid. It demarks Spring's original decklid height." There's a rumor that this change occurred by accident at Murray, where Jet bodies were built, but this is quite unlikely—Murray did not make that sort of mistake, its business was built on creating exact copies of production prototypes.

One thing is certain—Frank Spring was most unhappy with the production Hudson Jet. A slight family resemblance to the 1954 Hudsons was maintained in the front end, which one unhappy owner told Floyd Clymer "looks like a guppy's mouth." The rest bore no resemblance at all to the still relatively low Hornet and Wasp. "Barit," one Hudson executive commented, "ruled the roost. He would come over about every three weeks and would go over the drawings with Mr. Spring, or Mr. Spring would take illustrations over to Mr. Barit's office, and this is how the thing originated."

Millard Toncray engineered the Jet, and as with predecessor Step-downs, he built it like an armored truck. It tipped the scales at 2,700 pounds—slightly less than Ford-Chevy-Plymouth but well over its compact rivals: The Aero-Willys weighed 2,500 pounds, the Rambler 2,550, the Henry J 2,400. Still, the Jet was regarded as a sparkling performer; Hudson had seen to the provision of a potent powerplant with enough special options to make a goer out of it, if the buyer so stipulated.

Hudson called the Jet's engine "a scaled-down H-145," which it definitely wasn't, though many roadtest articles included this catch-phrase. A close look at the specifications would reveal the truth: With a three-inch bore and 4¾-inch stroke, the Jet's 202-cubic-inch six used the same bore and stroke of the old Commodore Eight—discontinued for 1953. Hudson, taking a dose of good old cost analysis, had chosen to save money by using existing Commodore Eight tooling, subtracting two cylinders and creating the long-stroke Jet six!

There was a subtle advantage to this seemingly retrograde act. The relatively small bore allowed the L-head engine slightly higher compression

Jets on the line, April 3, 1953.

than could be achieved by a squarer bore and stroke. John Bond figured that 7:1 was the absolute limit for a 'square' L-head, without a sacrifice in volumetric efficiency. The Jet managed 7.5:1 in standard tune, and 8.0:1 optionally. Also, Bond noted, "The piston speed factor (of a long-stroke engine, which bothered many non-automotive people like Consumer's Union) can be alleviated by good design and becomes negligible when used with an overdrive or a dual range Hydra-Matic. During [Jet] tests we found that valve-bounce occurred at an indicated 70 mph in second gear, 62.5 actual mph. This is 5500 rpm, equivalent to a piston speed of 4,350 ft. per minute! During the performance tests we approached these speeds at least *75 times*, and the little stroker was still running as smoothly and as

Jets were popular with police departments, such as this Peru, Indiana, organization. At right is D. C. Fair, a local Hudson dealer and at the left, R. A. Atitt, Peru's chief of police.

Dorothy Mignault and Clair Emory drove a Jet from New York to San Francisco in three days and eight hours during the summer of 1953, paring two days off the cross-country time of the 1916 Super Six.

quietly at the end as at the beginning." Obviously Hudson had given the L-head engine a new lease on life.

The array of engines, transmissions and rear axles available on the Jet was large. It was the only compact besides Rambler with a Hydra-Matic option: The GM dual driving range Hydra-Matic cost $176 extra. Overdrive was also available, at $102. Both Jet and Super Jet (the difference was mainly in the interiors) could be had with the standard 7.5:1 compression, single carburetor engine providing 104 bhp at 4000 rpm. A high-compression aluminum head and Twin H-Power were both available—not necessarily in combination. With the 8.0:1 compression aluminum head, the Jet produced 106 hp. With this head and Twin H-Power, it belted out 114 eager horses. With Twin H-Power alone (few were built this way), the figure was about 110 hp. But in any form the Jet had a definite, and sometimes remarkable, weight/horsepower advantage over its competitors:

104 hp Jet 26.0:1
114 hp Jet 23.6

Willys 6	28.4	Henry J 6	31.9
Chevrolet	30.0	Plymouth	31.7
Ford V-8	30.0	Studebaker Champ	33.6
Rambler	30.6	Willys 4	34.0
Ford 6	30.8	Henry J 4	35.7

The Jet buyer's control over his drivetrain was extended to the rear axle ratio, an area full of choices. Manual ratios were 4.10 (standard), 4.27 for mountainous territory and 3.31 for maximum economy. With overdrive the 4.27 was standard, but 4.10, 3.54 or 3.31 were available. The 3.54 was standard with Hydra-Matic, but a 3.31 was also available and sometimes recommended for best economy. These various axles had some influence on Jet performance, but whatever the combination, it was well ahead of the Rambler, Willys or Henry J. Performance figures recorded by road testers are as follows:

	0-60 (sec)	¼ mi (sec)	top speed (mph)
Popular Mechanics Super Jet, automatic	14.6	19.6	102.0
Autosport Review Super Jet, manual	16.6	20.5	88.5
Road & Track Super Jet, overdrive	15.1	20.1	91.1
Motor Trend Super Jet, automatic	15.2	19.8	102.4
Auto Age Super Jet, automatic	12.8	20.9	90.0
Speed Age Super Jet, automatic	12.5	n.a.	96.0
Average of above	14.5	20.2	95.0

Though the engine options could be had in either model, the factory and dealers kept the basic Jet in short supply, and it was available in four-door sedan form only. The Super Jet two- or four-door sedan was mainly distinguished by richer fabric upholstery, and a colorful if chromey dashboard with the usual warning lights for oil pressure and amperes. Hudson used an attractive two-tone worsted weave combined with broadcloth, and was not stingy with special touches like a rear compartment robe rail and a large tailored pocket built into the back of the front seat. While Hudson's description of Jet interiors (Salon Lounge) was going a bit too far, they were of good value and quality.

The Jet's introduction stole most of the thunder from the larger Hudsons for 1953, though in truth the face-lift was very mild. Strut bars were eliminated from the grilles—for the cleanest frontispiece the Big Hudsons had ever presented—and Super Wasps and Hornets (with Super Jets) used a fake airscoop on the hoods. The Pacemaker was dropped, and in its place appeared the Wasp, on the 119-inch wheelbase using the 232 cid engine, and the Super Wasp (also 119-inch w.b.) with 262 cubic inches. The three-passenger coupe was eliminated, but other body styles remained pretty much as before, and no new 1953 bodies appeared other than the Jet's:

Model Body Style (passengers)	Price	Weight (lb)
Series 1C Jet		
sedan, 4dr (6)	$1,858	2,650
Series 2C Super Jet		
sedan, 4dr (6)	1,954	2,700
club sedan (6)	1,933	2,695
Series 4C Wasp		
sedan, 4dr (6)	2,311	3,380
sedan, 2dr (6)	2,264	3,350
club coupe (6)	2,311	3,340
Series 5C Super Wasp		
sedan, 4dr (6)	2,466	3,480
sedan, 2dr (6)	2,413	3,460
club coupe (6)	2,466	3,455
Hollywood hardtop (6)	2,812	3,525
convertible brougham (6)	3,048	3,655
Series 7C Hornet		
sedan, 4dr (6)	2,769	3,570
club coupe (6)	2,742	3,530
Hollywood hardtop (6)	3,095	3,610
convertible brougham (6)	3,342	3,760

The 1953 Hornet, which came in the same four body styles as the year before, was offered in a wide range of colors, with two-tones done to

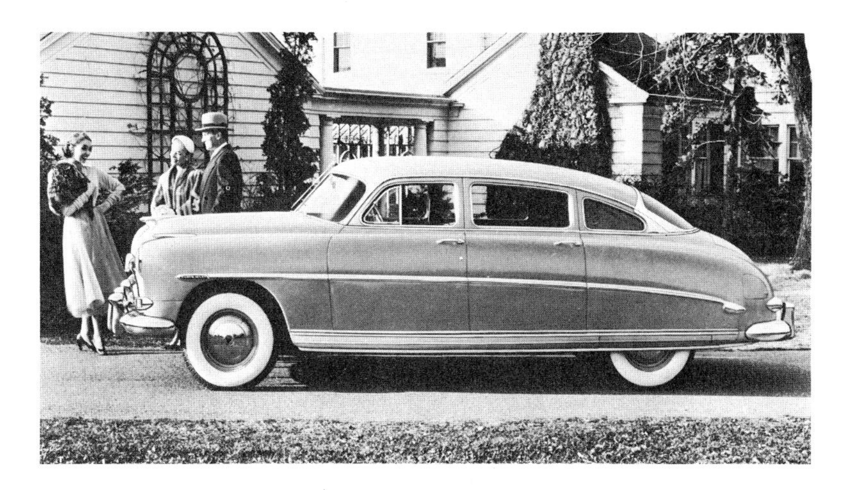

The 1953 Super Wasp.

resemble a hardtop on the club coupe model—the roof color was carried down to the beltline. The major alteration for 1953 was in its interior, which was fitted in new nylon check weave upholstery in combination with a sturdy vinyl that Hudson logically named Dura-fab. The new Super Wasp was really last year's Wasp series, while the latter designation was applied to the low-priced ex-Pacemaker range: The comparable Wasp was in fact the same price as the 1952 Pacemaker. The Wasp used all-cloth upholstery, in solids combined with a tan woven cord, with red and brown stripes; the Super Wasp was done up in Dura-fab combined with a less luxurious fabric than the Hornet. Although the brochures insisted that Hudson "sets the style in luxury, performance and durability with a sparkling new line," there was hardly a thing that could be said to be new. This is by no means a sin, but among Detroit cars in 1953, 'new features' translated into 'sales,' which Hudson could have used.

The big Step-downs were thus all the more easy to relegate to the background vis-à-vis the Jet, which netted the lion's share of publicity. Predictably, most testers found that this new car—the only all-new U.S. car for 1953—was quicker than its competition, solidly built, and eminently practical though unpretty. *Road & Track*, perenially ready to laugh at any small car from Detroit, admitted that its collective mind was completely changed after its Jet road test. It praised comfortable seating, good visibility and driving position, quality of fit and finish inside and out. "Demonstrate this car and it will sell itself," was John Bond's summation. Of course a car like the Jet was bound to be liked by people for whom an Austin A-35 was

The 1953 Hornet sedan. The Step-down's trunk, though small in appearance, had considerable capacity.

ideal family transport, and the foreign car crowd was duly exuberant. "You can cruise the Jet all day long at 70 to 75 mph without any feeling of engine strain," said John Bentley in *Motor Age* (you couldn't do *that* in an A-35!), "nor is it necessary to stand on the brake pedal . . .nine inch drums provide over 132 square inches of lining area. . . .Compare these figures with those of other Detroit products lugging from 500 to 1,000 pounds more. Congratulations, Hudson."

Auto Sport Review had about as accurate a summary of the Jet as one may find: "It is a solid, well-constructed, nimble car with distinctly lively performance, real lugging power and good handling qualities. The engine, despite its long stroke, should be long-lived and easy to service (it is quite accessible). The body, in one unit with the frame members and with more than 5,000 welds, is sturdy and shake-proof and promises a long, rattle-free life. The quality of the upholstery and trim is better than average, the chromium is top quality for this day and age and the paintwork has no orange peel. It is not 'a new kind of car' as the Hudson ads say; it is a reasonably scaled-down version of the conventional American automobile. It is much more maneuverable, has better acceleration and road holding than the average, may have a long trouble-free life and with overdrive, should be more economical."

Consumers Union, for whom the boxy, upright, simultaneously clumsy-looking but practical Jet must have been a revelation, was pleased. CU compared it with another favorite, the Willys Aero, and called both cars good buys while admitting that they'd both depreciate rapidly. It couldn't quite bring itself to say the Jet would be more durable than a Plymouth or a Ford (Plymouths were well-known durable goods in 1953), but CU had to say both the Willys Aero-Falcon and the Jet could "be recommended for many automobile buyers, both for their size and other valuable design features . . .with motorists who can use their advantages and can afford to pay for them." Rather backhanded praise, but in the traditional idiom of an organization that looks at automobiles as it does refrigerators.

In its haste to publicize the Jet, Hudson committed a rare breach of ethics that would leave hell to pay today: A sales booklet entitled *Quotes from the Experts* blithely edited-out anything questionable in 'reprints' from *Motor Trend, Speed Age, Auto Age* and the like. It's amusing to compare *Quotes from the Experts* with the original tests.

For instance, while the magazine saluted Hudson's heritage of roadibility, it pointed out that "the Jet has few of the characteristics of the road-hugging Hornet," and in a tight turn at 50 mph the inside front tire "seemed nearly off the ground." The car's stopping power was not up to other comparable models, taking 213 feet, for example, to stop from 60 mph while the competition averaged 189 to 205 feet.

Although *Motor Trend* found space utilization fair, there were bad marks for Jet comfort levels: "The shape of the seat does not lend itself to driver comfort . . .usefulness will be limited to short trips with a full passenger load." As for ride, above 50 mph "oscillation is excessive. Bottoming, accompanied by three or four oscillations, occurred at 70 mph, and the front end felt nearly airborne. On sharp dips, full spring travel was reached at 50, and on very severe dips the Jet bottomed at 30 mph."

Testers also concluded that the windshield wipers did not clear an adequate portion of glass, and that certain design features were unfortunate. The spare tire, for instance, had to be lifted to clear a six-inch sill at the rear of the deck, and though the driver could see the right front fender easily enough, "he won't see the right rear fender without getting out of the car.

"To be a real threat," *Motor Trend* summarized, "the Jet will need more than its excellent performance to make that long, uphill grind to the top of the low-price field. Its engine needs few mechanical improvements, and its body lines, typical of its class, need no period of introduction to the car buyers. A wider [track] plus heavier shocks would improve stability. Increased rear passenger space would make the Jet a stronger challenger in its chosen price class.

"The next two years," *MT* concluded (and Hudson omitted), "will tell whether buyers will respond en masse to its features, or whether they prefer a more conventional package." In fact it would take hardly more than a year.

With a fast start (not for the Jet, but for the big cars), Hudson built 65.7 percent of its 1953 output during January-June, and recorded a small profit in the first quarter. But continued parts and material shortages, plus wildcat strikes and delays in Jet production held volume to almost exactly that of 1952: 76,331 cars. In a growing year for the industry as a whole, Hudson's market share dropped from 1.76 to 1.24 percent, and it fell to fifteenth in the production race (seventh among auto companies)—being passed by Packard, which was responding to its hard driving new president, James Nance. Heavy expenses were incurred through operating losses, the Jet tooling program, increased inventories and new defense commitments. As a result Hudson experienced a goodly loss—$10,411,060 on sales of $193 million. It was the firm's biggest loss in history.

Ed Barit explained: "The loss for 1953 was caused by numerous problems—inability of dealers to handle the anticipated volume of cars due to a combination of wholesale credit restrictions imposed by finance companies and overstocked new and used car markets. [The seller's market had screechingly halted.] Serious delays occurred in the introduction of [the Jet] because of steel shortages, die problems and a strike in the entire die shop industry."

In 1953, it was a good practice to pull over if this appeared in one's mirror.

The Jet had cost $12 million to tool, and Hudson had to pay for it. When the shortages arose there were repercussions among Hudson creditors. Murray was supplying Jet bodies—the first time an outside firm had built Hudson shells since the early thirties—on a delayed payment plan. Explains Roy Chapin, "As I recall, Murray tooled quite a lot of [the Jet] by amortizing by a charge of say $50 a body, thereby saving the manufacturer the initial capital outlay. As I can remember, there were serious hassles with Murray that the projected schedules were not met, and consequently Hudson amortization and recovering of its tooling costs were running way behind what it had anticipated." This helps explain the apparent anomaly by which Hudson could build 80,000 cars and make $8 million in 1952, yet lose $10 million on almost the same production in 1953. "Take a look at the amortization schedules," Chapin says, "because there was a period there [when] I can remember some banker friends saying 'You're not really amortizing those tools,' which of course were charged against earnings. In an automobile company your depreciation and amortization policy can materially affect, on a yearly basis, at least your bottom line figure." Barit reported at the time that as a result of sales

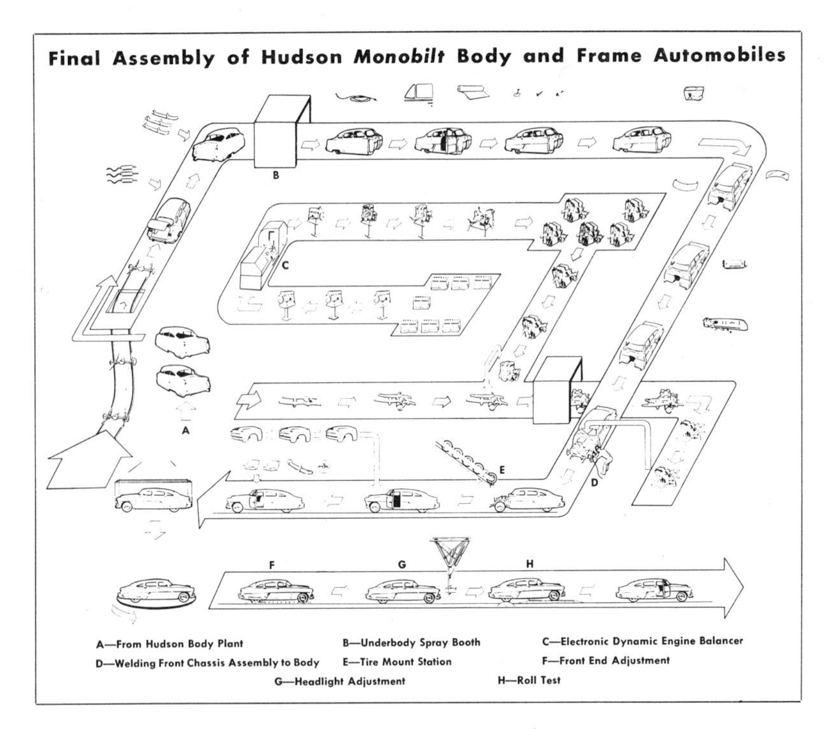

Final assembly diagram.

conditions, "we found it necessary at the close of the year to review and adjust all of our amortization charges for tooling and other items incidental to model life." Hence the quarrel with Murray. Both sides were unhappy—Hudson because Murray wasn't building bodies fast enough, Murray because Hudson wasn't selling enough of them anyway.

Production was held up because of delays in the supply of Hydra-Matic transmissions. GM's output of Hydra-Matics was interrupted in 1953 by a plant fire (which resulted in the shift of Hydra-Matic to Kaiser's vacated Willow Run Plant) and a number of strikes. Hudson's 1953 model year count was 66,143: 21,143 Jets, 17,792 Wasps and 27,208 Hornets. Defense was a bright spot—Hudson had contracted with the Martin Company to supply air frame components for the B-57, and continued to deliver parts of B-47's to the Wright Corporation.

Hudson was not the only firm in trouble, of course. In 1953 Kaiser Motors and its subsidiary Willys Motors lost $15.3 million in the first nine months. Studebaker had not done particularly well. Packard had rallied slightly, but plunged again toward the end of the year. These gloomy developments for the independent manufacturers were measurably magnified by trouble inadvertently caused by one of the giants. In the industry, men refer to it as the 'Ford Blitz.' Says Roy Chapin with a wry expression: "It's just wonderful, all the things we've lived through."

Henry Ford II, who became president of Ford after the death of his grandfather, had largely left the running of the company to Ernie Breech and a handful of well-trained, competent executives. By 1953, though, Breech had pulled the company out of a rather perilous postwar position, and young Henry began to take more of a role in charting its course. Determined to make Ford number one again, "or break it in the process" as one commentator put it, they began shipping Fords to dealers whether they were ordered or not. To keep the inventories down, dealers resorted to massive discounting—some said they were even taking losses—and General Motors soon reacted in kind. But the independents were not able to take this kind of a loss on their products, and neither were their dealers. They—and Chrysler too for that matter—had no effective response to this battle of the giants.

By the end of 1954, Ford had moved from about 300,000 units behind Chevrolet to just 17,000 short. But Chevrolet had had a banner year on its own, with 1.4 million cars. Neither was hurt, but the independents (and Chrysler) were reeling. Buick and Oldsmobile shot past Plymouth, traditionally number three in industry production; Cadillac moved into the top ten, and there wasn't an independent higher than eleventh. Sales of Hudson, Packard, Kaiser-Willys, Nash and Studebaker all plunged downward. By 1955 there wasn't an independent left which hadn't merged with another independent, and Chrysler had borrowed $100 million to completely retool and do battle on more even terms.

One of the hoariest of old chestnuts is the tale bandied about by enthusiasts of Hudson, Studebaker, Packard and Kaiser-Frazer, that the Big Three somehow conspired to do in the independents. It never was true. The large corporations many times aided independents with transfusions of talent, equipment and advice—GM, indeed, provided Kaiser with some steel stamping work when their Willow Run plant was largely idled during the sales slump of early 1951, and GM Hydro-Matics were sold willingly to Nash, Kaiser-Frazer and Hudson. But the independents were undoubtedly casualties of the great wars between the giants, and even Chrysler was reduced after 1954 to a sort of second class power status—some people began referring to 'the Big Two-and-a- Half.' So if Ford and GM are to be blamed for anything, they stand easily convicted for staging a price war that hurt not each other but the small manufacturers who together didn't account for more than ten percent of the business. And they have paid for such precipitate action in the years since, being constantly scrutinized by antitrust bureaucrats, looked upon as bloated oligopolists to the point where they can't win an argument over safety or fuel economy, even when they're right. It can be fairly argued that this all-or-nothing sales drive marked the beginning of distrust of automakers by government and some segments of the retail sales industry. In the long run the Ford Blitz hurt Ford as much as anybody.

With the price war between Ford and Chevrolet at full pitch in late '53, the 1954 Hudson introduction was almost anticlimactic, and nobody much bothered to note that once again for the seventh straight year, Hudson was offering its original 1948 Step-down design. The '54 Jet, needless to say, wasn't changed a bit.

Model Body Style (passengers)	Price	Weight (lb)
Series 1D Jet		
sedan, 4dr (6)	$1,858	2,675
utility sedan, 2dr (6)	1,837	2,715
Family Club Sedan (6)	1,621	2,635
Series 2D Super Jet		
sedan, 4dr (6)	1,954	2,725
club sedan (6)	1,933	2,710
Series 3D Jet-Liner		
sedan, 4dr (6)	2,057	2,760
club sedan (6)	2,046	2,740
Series 4D Wasp		
sedan, 4dr (6)	2,256	3,440
club sedan (6)	2,209	3,375
club coupe (6)	2,256	3,360
Series 5D Super Wasp		
sedan, 4dr (6)	2,466	3,525
club sedan (6)	2,413	3,490
club coupe (6)	2,466	2,475
Hollywood hardtop (6)	2,704	3,570
covertible brougham (6)	3,004	3,680
Series 6D Hornet Special		
sedan, 4dr (6)	2,619	3,560
club sedan (6)	2,571	3,515
club coupe (6)	2,619	3,505
Series 7D Hornet		
sedan, 4dr (6)	2,769	3,620
club coupe (6)	2,742	3,570
Hollywood hardtop (6)	2,988	3,655
convertible brougham (6)	3,288	3,800

Hudson's 'Flight Line Styling' for 1954 was really a fine face-lift to the Hornet and Wasp. Front ends were cleaned up dramatically with a simple, pleasing grille of stainless steel and chrome-plated white metal; a single center bar bore the Hudson emblem. Overall, the frontal design gave an impression of low build and wide stance. The sides were cleaner and, if not really different, improved by careful ornamentation. Huge triangular taillights, visible from the sides as well as the rear, were set high into the rear fenders, which had been raised enough to allow the driver to see them for the first time on a Step-down. The trunk lid too had been raised to impart a more squared-off look to the rear end, and thinner trunklid hinges were less of a threat to luggage. Massive bumpers wrapped around the edges of the rear fenders; a built-in license plate frame in the rear bumper formed a center guard. Inside, there was a new, much more modern and attractive instrument panel, which consigned the gauges to a shrouded housing, their dials illuminated by indirect lighting. ("The 'antique shop look' is gone," said *Motor Trend*.) The new '54's used one-piece windshields, and rear quarter windows were elongated for extra visibility.

Super Wasp Hollywood hardtop, 1954.

Mechanically, Hudson offered Instant Action engines with Super Induction, which translated to 7-X engine modifications—particularly relieved blocks (the area between piston and valves was scooped out, permitting a more efficient passage of intake mixture and exhaust gases), hotter camshaft and 7.5:1 aluminum cylinder head. The stock Hornet now offered 160 bhp, and 170 with Twin H-Power, an $86 option. Power brakes and steering were optional on Hornet, Super Wasp and Wasp.

Hudson power steering was the direct-action linkage type, activated when the steering wheel was turned as little as two degrees and provided

Trunk lids of the big Hudsons for '54 were redesigned for added space. Thin hinges were used, along with three-way-visible taillights. The 1954 Hornet club coupe with continental accessory; also available on the Super Wasps.

Hornet convertible brougham 1954.

up to eighty percent of the required steering effort. Since Hudson used the same steering ratio for both the power and standard setups, it took no extra effort to turn the wheels should the power booster fail. "Spring resistance of four pounds gives a constant feel of the road and prevents dangerous oversteering. Hudson Power Steering works only when needed with the car remaining under conventional manual control until more than 'fingertip' steering effort is required. After a turn the car automatically resumes a straight course," said press releases. *Motor Trend* didn't buy it: "[It] is more responsive to wheel movement, but the long-favored feeling of being master of the man-sized Hornet is missing. . . .Because of Hudson's originally fine steering, and because this is their first car with power steering, we're inclined to think that our criticisms may not apply to later models."

The magazine didn't care for the power brakes either, saying they failed to pull up a Hudson as well as the standard 1953 brakes: "After repeated stops, the brakes faded badly and grew noisy." They were also

The 1954 Hornet typified new styling with its clean grille and hood scoop. The 1954 Hornet sedan.

Interior of 1954 Hornet Hollywood featured Plasti-Hide upholstery in antique white with choice of red, green or blue trim, color-keyed instrument panel and Plasti-Hide headliner.

very sensitive. But Hudson's traditional reserve braking systems made them among the safest power brakes on the market: if the power source failed, a reserve vacuum tank permitted up to three power applications; direct mechanical linkage also operated brakes hydraulically, without power; and even this reserve had a reserve, a second hydraulic fluid tank to keep it filled to a safe level at all times. Finally, there was the mechanical handbrake. A Hudson driver didn't need to fear getting his car to stop.

The Commodore Six which had ranked just below the Hornet in 1952 was not replaced in 1953. But in mid-1954 the slot was filled by a new Hornet Special, priced about $150 below the Hornet but running the same wheelbase and engine, with the Twin H-Power 170 horsepower option. Available in four-door, two-door and club coupe, the Special was, on paper, the hottest of Hornets. The club coupe weighed only 3,505 pounds, making it the lightest Hudson yet with the big 308 engine. Hornet Specials offered solid luxury interiors, with check-weave worsted fabrics in a choice of blue or green to harmonize with exterior colors, contrasting bolster cloth and Plasti-Hide trim.

Both the Wasp and Super Wasp used the shorter 119-inch wheelbase, and were distinguished from each other by interior trim and powerplants: again 232 cubic inches for the Wasp, 262 for the Super Wasp. With standard head, the Super Wasp engine developed 140 bhp at 4000 rpm; the optional 7.5:1 aluminum head gave it 143 bhp; Twin H-Power made it 149. The Wasp engine developed 126 horsepower, or 129 with the special

The 1954 Jet Liner.

The 1954 Super Jet.

head; it was not available with Twin H-Power. Twelve solid colors and twelve two-tone combinations were offered on the big Hudsons, though the Wasp was not available in the latter.

The 1954 Hudson Jet was virtually identical to the 1953 version, though its grille had small raised bumps for quick identification. The lowest-priced series 1D line was supplemented by a two-door utility model, containing a practical drop-down rear seat that allowed a healthy cargo bed, extending the trunk into the rear compartment. Hudson had also responded to complaints about limited leg room in Jets, by moving the rear seat two inches further back.

A brand-new series was the model 3D Jet-Liner, luxury version of Hudson's little car and beautifully fitted with a quality interior. Embellished on the outside by chrome trim for the windows, gravel shields and door-base moldings, the Jet-Liner featured an interior of pleated vinyl over foam rubber cushions, its headliners done in Antique white Plasti-Hide. Contrasting colors of red, blue or green were offered in vinyl-trimmed doors, and instrument panels were appropriately color keyed. (Super Jets used blue or green worsted cloth seats and vinyl door panels, plain-Jane Jets were correspondingly more cheaply upholstered.) The wide range of Jet drivetrain options was retained.

Hudson had manufactured some 'severe usage' parts for the Jet in both 1953 and 1954. These seemed genuinely designed for heavy-duty or police work (some departments were enthusiastic Jet backers) rather than racing. Among the items offered were front anti-sway bars, heavy-duty shocks, drag-race-type rear axle units (3.92, 4.09, 4.27 and 4.89), hotter camshafts and brake assemblies from the larger Hudsons.

Hudson shipped only 32,287 cars during 1954, but those who bought them liked them well enough. "Hudson offers more safety features than any other car," an El Paso salesman told Floyd Clymer during a public survey. "Triple-safe brakes should be compulsory on all cars," a Florida realtor echoed. "I never slow down for curves," said a Kenmore, New York mechanic, "no sway on curves at all." In general, people still found Hudson roadability excellent, its performance more than adequate—and its trade-in value poor. "If you buy a Hudson," one testing publication put it, "plan to keep it a long, long time." Replied a Michigan businessman, "If people who buy cars from habit would compare and drive a Hudson, there would be a lot more Hudsons sold."

Motor Trend's criticisms aside, the magazine still liked the Step-down, and its comments strongly echoed the views of owners. "The Hornet's records in stock-car racing are due partly to its outstanding roadability," *MT* wrote in March, 1954. "The Hornet is not to be denied praise as one of our top road-hugging machines; rutted, bumpy roads meant nothing to MT's test car, and even at high speeds on rough roads, the rear end remained stable. There was no steering wheel vibration or shock, and no wheel hop at any time during the test. . . .Rather than go to a new powerplant for more performance, Hudson continues to increase the power in the big six-cylinder engine. With Twin H-Power, the 308-cubic-inch engine does better than many of its competitors. In figures, this power means a 0-60 time of 14.8 seconds, a standing quarter-mile time of 19.3 seconds, and an average top speed of 100.34 miles per hour. By comparison, MT's '53 Hornet test car (160 bhp with Twin H-Power) turned 0-60 in 15.5 seconds, went

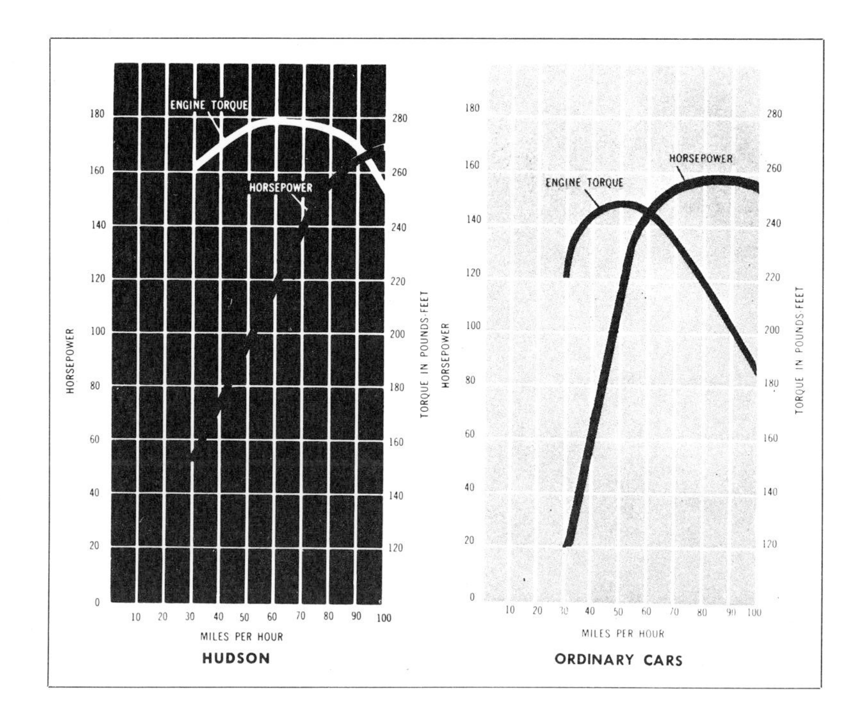

Torque-horsepower comparisons from the 1954 Hudson sales brochure.

Jet Utility sedan converted to a carryall with removal of the rear seat, as trunk divider wall fell flat.

through the quarter-mile traps in 20 seconds, and averaged 98 mph. This year's car showed an increase in gas mileage over the '53. The '54 Hornet's fuel consumption in our constant-speed and traffic checks averaged 18.08 miles per gallon; the '53 test car averaged 15.8. So Hudson can proudly claim a 13 percent increase in economy, along with power and performance gains. A higher compression ratio, Super Induction, and other engine changes are responsible. High-speed fuel consumption was about equal in the two cars, but the '54 showed a noticeable increase in traffic economy. It is about average in its speed."

Summing up, *Motor Trend* rhetorically asked itself "What will I get for my money if I buy a Hudson Hornet?" and replied, "Power and quality. The engine will provide pulling power to match any need, and it has an adequate safety margin for passing at high speeds, yet has family-car docility. The drawback to this is bothersome engine noise during low-gear acceleration.

"The Hornet has exceptional roadability; there is little chance of a minor driver error getting the car and its passengers into real trouble on a fast turn. Yet anyone used to handling past Hornets will notice changes with the power steering. A fine highway car, the Hornet may seem bulky at first in city driving; its power accessories will eliminate any lasting problems . . . The well-built Hornet will give roomy, comfortable travel, reasonable economy, and above-average performance—at a price below the high-priced cars. The potential buyer—the buyer who shops for the top buy materially, not for the best buy trade-in-wise—will be impressed with the Hornet's true value."

Wilbur Shaw tested a 1954 Hornet for *Popular Science*, and threw a few hard punches on behalf of the L-head engine. He led off with some interesting comparisions of horsepower per cubic inch on competing overhead valve models:

Horsepower per Cubic Inch

Lincoln V-8	.64	Oldsmobile V-8	.54
Cadillac V-8	.63	Studebaker V-8	.51
Hudson six	.55	Chevrolet six	.49
Buick V-8	.55	Ford six	.47
Chrysler V-8	.54		

The 1954 Hornet Special club sedan featured more sweeping lines than the Hornet-Wasp club coupes. A Hornet Special club coupe was also available.

Hornet Special interiors were done in check-weave worsted fabrics of blue or green, and Plasti-Hide bolsters. The seats were 64 inches wide.

"On the strength of the figures," said Shaw, "I'd say some of us have been selling the L-head down the river. [And essentially,] all that Hudson has done to its '54 Hornet engine [is] made it breathe better. The gas gets in and out of the cylinders more easily. . . .These little changes have had happy results beyond the boost in horsepower. The car's faster. I sneaked a look at the confidential results on one engineering test car that showed a top speed in excess of 115 mph. The torque peak . . .is practically constant now all the way from 60 to 80 mph. [Hornet torque with the 170 hp engine was 279 foot-pounds at 2500 rpm.]

"The L-head engines had qualities inherent in the design that the overhead valve has to contrive. The L-head, as everyone knows, has valves that work up and down in an antechamber off each cylinder. The incoming air/gas charge has to turn corners to get into the cylinder. The burned gas has to turn corners to get out.

"Champions of the overhead valve engine insist the ohv is more efficient. The gas doesn't have to turn all those corners getting in and out. What they don't say is that the corners of the L-head provide a natural turbulence in the incoming charge. The gas molecules tumble and swirl like sawdust coming off the blade of a table saw. And proper turbulence is the main reason an air/gas charge burns rapidly and completely in a cylinder, particularly at high speed.

"It wouldn't pay to quarrel with the ohv people. They've got something. The steadily increasing number of overhead valve engines in U.S. automobiles is testimony to that. But the case for the overhead valve isn't cut-and-dried. The ohv mechanism is mechanically complicated. One

typical ohv six has upward of 120 more parts in its valve system than the Hornet's.

"In the course of my visit to Hudson in Detroit, I remarked to one of the company engineers that automobile racing engines invariably have overhead valves. I was pretty sure what his reply to that would be: 'Sure, but they have overhead camshafts too. Mechanically, the overhead valve and cam—from the cam face onward—is as simple as the valve system on the L-head.'

"He could also have added that the speed at which racing engines are run assures proper turbulence in the combustion chambers."

Hornet Special engine with optional Twin H-Power delivered 170 hp.

The 1954 Hornet Special sedan.

Like *Motor Trend*, Shaw complained of the '54's brake fade, but he was sure an improvement would come there. "It has to. And for the Hornet, I'll bank for a solution to the problem on a conservative company that plods along making better L-head engines that turn on with a roundheaded key."

But time was running out.

The last new Hudson-built model, was introduced on April 11, 1954—the Jet Family Club Sedan, with an announced price of $1,465. This figure was lower than any of the Big Three, though it soon rose to around $1,600. According to N. K. VanDerZee the Family Club was designed to penetrate new market areas, populated by "millions of American families without an automobile, and many more in need of an economical and dependable second family car. By careful planning, we were able to effect new economies in the production of a Jet Family Club Sedan which will bring Hudson quality and performance within the reach of many more families." The truth was starker than that. Bluntly, Hudson dealers couldn't give away the awkward looking little Jets priced several hundred dollars

Smart interior of the 1954 Jet Family Club Sedan featured striped upholstery, color-harmonized with solid bolster cloth. The seats were almost as wide as the car was high. The Family Club Sedan.

above Big Three competition of 'standard size' and had to, somehow, move them out. (One Jet convertible was also built, a red model, which AMC-Hudson sales manager Virgil Boyd bought for his son.)

The Family Club weighed in at 2,635 and was stripped to the bone—no brightwork on the outside, cheap cloth upholstery on the inside. Available in a limited choice of six colors, it offered the usual wide range of drivetrains, but failed to make a dent in the gathering deficits. Through April 30, 1954, when Hudson closed the books as an independent company, it had lost $6,231,570 on sales of only $28.7 million. Had the year continued at that pace, Hudson would have lost $18.7 million, almost double its greatest loss in history.

By then, of course, it was already official that the Hudson Motor Car Company would merge with the Nash-Kelvinator Corporation to form American Motors. Word of merger talks between Ed Barit and Nash's George Mason had leaked as early as October 6, 1953, just four days after the 1954 model introduction. (This couldn't have done sales much good; word was out early that Hudson might be 'orphaned' by the deal.) The talks

Last of a distinguished line, final Step-downs shipped December 29, 1954.

were confirmed on November 27, and early in 1954 both companies made the coming integration official.

On May 1, 1954, the Hudson Motor Car Company ceased to exist. Henceforth the destiny of a marque that had been on the American scene since 1909 would rest on the corporate decisions of American Motors. And in those days AMC was perceived as a partnership of circumstance—like Studebaker-Packard and Kaiser-Willys born of necessity, and facing what appeared to be a dim future.

It is always easy, and usually not very precise, to single out a product as the cause of a company's decline. In Hudson's case it is particularly easy. Though its rapid plunge into fiscal disorder after 1952 was the result of many complex factors, the Hudson Jet was undoubtedly Hudson's single greatest mistake. It was the wrong car at the wrong time, an expensive white elephant that channeled off money which could have gone toward major updating of the big, powerful, well-built Hornets and Wasps for which Hudson had already won fame.

American Motors was now the future. But before we delve into Hudson's afterlife as a very junior partner to Nash, it is necessary to take a look at what might have evolved into a whole new generation of Step-down Hudsons—had the company stayed prosperous, and independent.

Italia and X-161: The Hudsons that Weren't

IN 1953, DWIGHT D. Eisenhower was taking office as president, the Korean armistice was being negotiated, and Detroit was talking sports cars. Some of the most unlikely sources produced two-seater predictions of the future. Cadillac and Oldsmobile, for example, traditional builders of big, heavy family cars, had truncated caricatures of their regular lines in the 1953 LeMans and Starfire showcars. A raft of sporty specials, constructed with production definitely in mind, were rolling out of Chrysler—equipped with sleek, Italian-built bodies. Chevrolet had announced its fiberglass Corvette, and Ford was promising a V-8 two-seater in the not-too-distant future. The independents were also active: Packard had built the Pan American in 1952, following with a production 1953 derivative called the Caribbean. In the same year Studebaker had a sports car in clay for possible 1955 production, while Nash was entering its third year with the attractive and rapid Nash-Healey. Kaiser-Frazer, down and out financially but still hoping for recovery, built the Darrin sports car. And Hudson, not to be outdone, created the Italia.

Though the Darrin had a fiberglass body and the Italia aluminum, these two cars had much in common. Both had sprung from enthusiastic designers' concepts of an elegant sports-type body on a workaday compact chassis—the Darrin's from a Henry J six, the Italia's from a Jet. Like the Darrin, the Italia bore no resemblance at all to the production car it was based on, and both cars were rolling laboratories of innovation from grille to taillights. Neither company made money on them, but the Italia was only six percent as plentiful as the Darrin: Kaiser-Frazer ultimately built 435 copies of its dream car, against just twenty-six Hudson Italias.

The man charged with selling the twenty-six examples was none other than AMC's present board chairman Roy D. Chapin, Jr. "I was the Italia sales manager," he says, "one of my more glorious roles in life. I got rid of them." When the author told Mr. Chapin that the Italia allegedly broke even for Hudson he replied, "Shows what a good salesman I was! Glad to hear you say it, but I don't believe it. The research and development figure for the Italia [$28,000] sounds like our racing budget in those days."

But there were a lot of good reasons for building this interesting limited-production car. With the competition displaying dream machines at every opportunity, a Hudson two-seater provided useful PR—a badly needed 'progressive' image for a company relying on a six-year-old design for its production bodies. In an age when predictions were rife, Italia was the Hudson prediction. Indeed, many state that Frank Spring's original Jet design used Italia features, particularly the fender-mounted brake-cooling air scoops. An Italian coachbuilder, Carrozzeria Touring of Milan, was pressing Hudson to let it do some special work, and its prices were attractive. Chrysler had already shown how useful an Italian connection (Ghia) could be from the standpoint of advertising and design studies. Mike Lamm has also suggested that the Italia program was a sort of 'reprieve' for Frank

The Hudson Italia at the Torino show, March 2, 1954.

Spring, who had been terribly disappointed with the Jet—an opportunity to again let his and his staff's imaginations run loose on Hudsons of the future.

"I think maybe a little bit of all those might have been true," Roy Chapin continues. "I know Frank was extremely pleased at being able to do the Italia, it was right up his line. . . .I think the other reason you haven't mentioned is that somebody got to Mr. Barit and convinced him that we had to have a more modern image, the old car was in production too long, we had to do something to dramatize Hudson."

The Italia *was* dramatic. Announced on August 25, 1953, it was a strong statement that the company was *not* locked into unchanging attitudes

Touring badge from an Italia.

toward its products, that it had many new ideas which, sooner or later, would appear in production. As Hudson put it, the Italia differed from most competitive show vehicles by being "a practical car with real production possibilities." John Bond of *Road & Track* echoed this sentiment: "Of all the current crop of dream cars, this one seems the most practical—an interesting combination of ideas." If this was an example of what Hudson would do in the future, admirers of the marque could be encouraged.

The Hudson Italia bristled with original, fresh thinking. On the road it sat nearly ten inches lower than a standard Hudson, while retaining the traditional Step-down design with its very low center of gravity. Its wheelbase (aside from the first prototype) was a practical 105 inches, its power derived from the hottest Jet engine of 114 bhp. Its windshield was wrapped around, putting Hudson dead-even with General Motors in that department. Front fender scoops gathered cooling air for the front brakes, fuselage intakes ducted more air to the rear brakes. The tail end was a little overdone, with a triple bank of chrome tubes which looked like exhausts but really housed tail-, directional- and back-up-lights, however the interior was quite revolutionary. The doors were cut fourteen inches into the roof for ease of entry; there were volumes of interior space (the Jet was exceptional in this area). Passengers sat on 'Anatomical' seats of clever design. Flow-through ventilation, the first in the industry, was provided via cowl air intake and exhaust grooves at the top of the rear windows—Frank Spring claimed that the air inside an Italia changed completely every ten minutes.

All the cars were upholstered in red and white leather and fitted with Borrani wire wheels; most were painted a light cream. Hudson preferred to call the Italia "a family car with undreamed-of styling, luxury and comfort" rather than a sports car, but Frank Spring possibly had the best definition when he said it was meant to carry two people and their luggage far and fast—in effect, the Italian definition of 'gran turismo.'

That the Italia was almost entirely Frank Spring's concept seems undeniable. Robert Barit, Ed's son, is given some credit for finalizing it, but Barit himself credits Spring for the overall concept while noting that the decisions to further develop and produce the car had to be corporate. Spring was certainly the motivating force. Says Chapin, "If I ever had to attribute it, it would be to Frank, because he was the only, what I'd call really forward-looking guy."

Carrozzeria Touring was an old firm which first won wide approval for its beautiful bodies on cars like Alfa Romeo in the thirties—the 6C *competizione* coupes were particularly admired—and later (with Farina) for influencing the proliferation of the integrated fenderline. Typical of small Italian coachworks, its premises were not impressive. Robert Barit told *Special-Interest Autos*, "I remember the Touring plant vaguely as a hole-in-the-wall operation down a narrow side street with a sort of production line snaking through a series of old dilapidated buildings, which perhaps had been built for the purpose. I remember they had high ceilings, and the plant wasn't far from the *autostrada*." Touring was one of the last old-line coachbuilders to give up the ghost, in 1967, following many successful *superleggera* bodies for Ferrari, Alfa, Lamborghini, Maserati and Aston Martin-Lagonda.

Touring's conceptual influence over the Italia was small; its value to Spring was its ability, like most Italian specialty houses, to build a number of custom bodies rapidly and get them to hang together well. Spring "admired the Italians particularly for styling," Stuart Baits said, "and this probably led him to Touring."

Most sources put the beginning of the Italia design program around the latter third of 1952, after the Jet had been finalized for production. In December of that year Lincoln swept the American stock class at the Carrera Panamericana, temporarily eclipsing Hudson's competition publicity, and this apparently spurred its development: Spring let on later that he had designed it with the famed Mexican road race in mind. It was his feeling that a lighter-bodied Jet with suitable engine mods, or possibly the Hornet engine, might provide an ideal combination to bring Hudson the same prominence in the PanAm as it enjoyed on the ovals of the AAA and NASCAR.

The Italia never got to Mexico, but it was well along toward public release when Barit alluded to it in a report to stockholders in the spring of

No.1 Italia had a metal grille and close-grouped letters.

No.2 Italia had die cast grille and bold, widely spaced letters.

1953: "A sports derivation of our new light car is being developed for possible limited production." It would be another year before Italias were actually available to purchasers, but by that time Spring had already shipped to Touring what was later referred to as a 'wheelbarrow': Jet front stub frame and suspension, floor pan and rear end components. Touring then reskinned this Jet foundation with the Italia body. The door hinge attachment points, for example, were stock Hudson Jet—the aluminum Italia-suit was simply fitted to the underpinnings. To make up somewhat for the loss of rigidity resulting from the aluminum panels and a tubular-frame upper structure, ¾-inch-diameter tubular braces were mounted the full width of the car, under the dashboard and behind the rear seat.

The Hudson-Touring relationship bears much resemblance to the Chrysler-Ghia arrangement put together by Virgil Exner, to transform Chrysler chassis into one-off for fifty-off specials. Ghia's first several efforts used stock chassis, but in 1953 it shortened a New Yorker wheelbase to 119 inches and a DeSoto to just 111 inches—creating respectively the Chrysler Special and DeSoto Adventurer. These cars looked a lot better than previous efforts on ten-foot wheelbases and, no doubt, interested Frank Spring. The Jet's wheelbase, of course, was already accommodating—just 105 inches—so no shortening was really necessary for a balanced look.

Italia No.1, however, did use a wheelbase of only one-hundred inches. Those that followed switched back to the standard 105. In addition, car No.1 offered other differences. It was leather-upholstered, while the 'production' Italias used combination leather and vinyl upholstery. Its dashboard was a custom layout including rev counter, oil-pressure gauge and an Alfa Romeo steering wheel, while later Italias used Jet instruments set into the cowl—though on all Italias the gloveboxes were central and the radio controls displaced to the far right. Car No. 1 was the only Italia fitted with overdrive—electrically operated. Its grille was die-cast while later Italias used aluminum louvers and bars, and its front fender contour was flatter than its successors'. All Italias used chrome-plated brass side trim, script, emblems, door handles, rear license plate frame, door/roof post covers; the oval grille shell and taillight 'pipes' were chromed steel. The 'production' cars also had chrome plated fender insert plates, which were absent on the No.1 car.

The Italia was shipped to the U.S. and underwent public introduction in the fall and winter of 1953. It was always a low-key thing, never with the impact of the Darrin—although the latter's designer, who went everywhere with it and often used a French accent to increase his allure, certainly had much to do with the good press Kaiser's dream car received. Italias were also shown in Europe. Reactions were moderately good. The car certainly

The 'production' 1954 Italia. Ample headroom for entrance and egress was provided by Italia doors, which cut into the roof by 14 inches.

stood very low, looked modern for the time, though its clean, oval grille was disrupted somewhat by its inverted triangle bumper arrangement—Spring's idea for Hudson identity. Said *Road & Track*, "The immediate reaction from the student of Italian styling is, 'Take off those ___ ___ pipes,' but the rest of the car shows an interesting combination of ideas. Outstanding features are the clean frontal treatment, Abarth-type fenders, brake cooling scoops, Borrani wheels, wrap-around windshield, Murphy doors [Spring had used the same door design for a Peerless while he was at Murphy] and generous interior space for two of the most luxurious individual seats imaginable."

The Italia's seats were very special, and were never duplicated by any other manufacturer. Twin buckets, adjustable for front-back position and rake, they each had two separate backrests. Strother MacMinn, a stylist who worked at Hudson for a time, says Spring "used different densities of foam for the bolsters. He varied density to give better support. He made a fair amount of human factor studies on the thing . . ."

Spring designed the seatbacks to give firmer support at the lower back than the upper, which is the anatomically correct method. Between the two cushions was air space for breathing, and the upholstery itself contained vent holes. Passenger movement thus caused the seats to 'breathe' through a sort of automatic pumping motion. Leather seatbelts were standard—Hudson was ahead of Ford in this respect, behind Nash, and tied with the Kaiser-Darrin—but they were attached to the seats rather than the body and thus of questionable value in a collision. Behind the seats was an immense cargo area with luggage tie-down straps and stainless-steel protective cargo-ribs.

On the strength of car No.1's reception, management decided to order twenty-five 'production' Italias from Touring, and on September 23, 1953 advised dealers that they could take orders for them at $4,800 per copy, or $4,350 p.o.e. New York. Nineteen orders were received—which isn't as bad as it seems considering the limited exposure the car had had and the fact that most customers were unable to see the thing in advance. The nineteen orders have caused some to state that only twenty Italias (counting car No.1) were built. But since car No.25 has been located and there is no sign of skipping, one can reasonably conclude that twenty-five, or twenty-six counting No.1, were built. Factory evidence is that the 1954 registered models were serialized IT-100001 through 100009, the 1955's from IT-1000010 to 1000025. The first car was not serialized. As Stuart Baits said, "If you're going to make two, you might as well go ahead and make twenty-five. If you tool up for it once, the first one will cost you maybe $150,000. The second might be $20,000 and the third $5,000 and so on."

Hudson collector and enthusiast Bruce Mori-Kubo fails to understand how only nineteen orders could have been taken. "At the Minneapolis Auto Show the Italia was a sensation," he remarked to the author. "Many people were attempting to place orders. I made every effort to place an order myself, and offered a very substantial deposit on a 'no return' basis, but could not get an order accepted. I was also told I could have the 7-X Hornet engine at 'extra cost.' I don't recall how much, but I was given a figure. I wanted to obtain that too but was unable to place my order. Many other people were in the same boat.

"I was a sales manager for Datsun back in the late sixties and early seventies. We had the same problem then on the 240Z, though we took orders with an unspecified delivery date. I would have given my eye teeth for an Italia and made every effort to place an order, but no go."

Although V-8's were all the rage when the Italia appeared, the fact that it had a six did not in itself hurt the car, as a two-seater special competing against the Corvette, Darrin and Nash-Healey. All these had sixes too. The V-8 Thunderbird had yet to be heard from. Baits, incidentally, mentions "some talk about using the Hornet engine" in the Italia, as Mori-Kubo had suggested. "I don't remember exactly why that wasn't done," Baits says.

The real sales problem was that the Italia was no quicker than the Jet. Its curb weight of 2,710 pounds (2,725 according to the Italia manual supplement) was exactly the same as the Super Jet club sedan. But Frank Spring told his friend Vic Racz that the 'prototype' Italia No.1 weighed 400 pounds *less* than that! Did Hudson Engineering, whose motto was diametrically opposed to Raymond Loewy's 'weight is the enemy,' insist on loading the Italia with ponderous reinforcement? Possibly. The engineers were very down on the Italia. "When you drove it you'd think it was just going to fall apart," Baits said. "The Jet was unit construction, and the strength of the body was in the body itself—the whole cowl structure was very strong, the posts were stiff. Our chief engineer [Toncray] was a nut on having cars solid and making them rigid. When you put a lot of aluminum panels in . . ."

At 2,700-odd pounds, the Italia's performance was on par with the Kaiser-Darrin, but there was a clear demarcation between Darrin/Italia and Corvette/Nash-Healey. Assuming our averages for Jet performance can be applied to the Italia with the same weight and engine, the following table is significant:

Model	Italia	Darrin	Corvette	Nash-Healey
list price	$4,800	$3,668	$3,523	$4,721
engine (dis.)	s-v 6 (202)	ohv 6 (161)	ohv 6 (236)	ohv 6 (253)
bhp @ rpm	114 @ 4000	90 @ 4200	150 @ 4200	140 @ 4000
transmission	3 sp.	3 sp. O/D	Powerglide	3 sp. O/D
rear axle ratio	4.1:1	4.55:1	3.55:1	4.1:1
wheelbase	105 in.	100 in.	102 in.	108 in.
length	183 in.	184 in.	167 in.	181 in.
weight	2,710 lb	2,175 lb	2,705 lb	2,990 lb
0-60 mph, avg.	14.5 sec	15.0 sec	11.0 sec	10.0 sec
top speed	95 mph	98 mph	106 mph	105 mph
lb/hp	23.7	24.2	18.0	21.4

Performance based on the following references: for Darrin, an average of *Motor Age* and *Motorsport* road tests; for Corvette, from *Corvette* by Karl Ludvigsen; for Nash Healey, from *Motor Trend*, September 1951; Italia figures are based on an average of Hudson Jet road tests, as an Italia was never tested by any publication.

Roy Chapin remembers that one of his Italia customers was bitterly disappointed in the car's performance: "He set out [in a brand new Italia] to go to northern Michigan. He had a Nash-Healey, and was a great friend of George Mason's—he turned around and came back, threw the keys on the desk and said 'Keep the goddamn thing, it won't get out of its own way.' It had the Jet engine, and whatever it weighed it was more than it should have."

George Mason's friend had a point. But this writer has driven the original condition Italia owned by Ray Robinson of Connecticut, and is obliged to defend a car with many positive attributes. In his opinion it has all the basic ingredients of a fine grand touring car *except* extreme straight-line performance. It steers lightly, corners flat—much flatter than any Jet he has sampled, evidencing different suspension settings. (Vic Racz says it used different spring rates). Despite what some have said about loose construction, the author found the Italia tight as a drum and beautifully assembled—which, it is interesting to note, was exactly how another writer judged a different Italia in California. The car's power was certainly not up to Corvette or Nash-Healey standards, but this could have been accomplished with the Hornet engine—the engineers say it would fit—and probably would have been installed had the car gone into production. The Italia interior is spacious and luxurious, with visibility approaching modern standards. The seats are excellent—their shape and design is obviously less an advertising gimmick than just good sound thinking. Italia styling doesn't much impress the writer, but there were a lot worse things around in 1954-55, and really it's just a matter of details, like the glittery rear pipes, that bother anyone who likes clean design. The overall shape of the car is a pleasing one, and it is a lot of fun to drive.

Unfortunately the car wasn't pleasing to conservative Hudson management, some of whom, even today, denounce it as "an advertising gimmick, with no production potential." It was held by Engineering to be totally lacking in production potential; some managers insisted that rain would get in past the cut-in doors, despite the fact that Spring (and today's owners) proved it would be carried off by gutters. Aside from this specific opposition, there was much debate about the car's styling as the basis for a whole line of passenger Hudsons to follow. At the retail level it was very expensive—Hudson dealers, remember, were selling Hornets that year for only $2,800—and the company never promoted it enough to sell in decent quantities. Whatever the faults of the Kaiser-Darrin—and it too was disliked by management and many dealers—Kaiser at least had the presence of mind to build enough of them to put into showrooms around the country, helping improve dealer floor traffic.

Carrozzeria Touring never planned to supply replacement parts for this sort of production, and they soon became a problem for the Italias in the field. In August 1956, Touring flatly refused to meet standing replacement orders and advised American Motors that no further Italia bits would be forthcoming. A. F. Buege of AMC's Milwaukee Parts Plant circulated a memo on August 14, noting that tail/directional light and front parking light lenses, deck and hood medallions, hood and fender nameplates were all needed and new sources for them would be necessary. Buege's boss J. A. Stratton added, "We have no idea what these parts look like and could not find a source unless a car is located that can be examined or a sample submitted by the ordering dealer." Evidently the Italians didn't realize the extent to which American manufacturers are held nose-to-grindstone for replacement parts by dealers and customers. But it did seem like an awful lot of bother for twenty-six automobiles.

By that time, of course, any possibility of the Italia being continued had been erased by the Hudson merger with Nash-Kelvinator. Hudson had announced Italias for sale in early 1954, AMC announced them again in August of that year. But after Roy Chapin had gotten rid of his two dozen, that was the end. Neither Hudson nor AMC ever printed literature, owners manuals or shop manual supplements for the car. The only piece of specific Italia 'literature' is a five page supplement to the Jet owners manual, which was tossed into Italia gloveboxes, almost as an afterthought.

Given the Italia's reception, it seems almost incomprehensible that it ever had a chance of being the basis for a new line of Hudsons. The evidence is strong, however, that it would have figured in a full restyling program had Hudson remained independent into 1955-56.

While the twenty-five Italias were being built in Italy, Frank Spring was back at work at Jefferson Avenue with the next logical step—a full-size, 124-inch wheelbase four-door version of the Italia called the X-161. In Hudson parlance the X was for 'experimental,' the 161 for the 161st prototype (perhaps since Spring had arrived). Aside from its larger dimensions, it completely reflected Italia styling, interior, relative lowness and lightness compared to production 1954 Hudsons. Like the Italia, it was aluminum-skinned by Touring. And like the Italia, it never had a chance.

Trever Constable limned the X-161 story in the September 1961 issue of *Car Life*, and because no one has significantly added to it since, it is best to quote his remarks: "One of the first 1954 sedans made in Detroit was taken by Hudson to Italy, complete with all accessories. At Carrozzeria Touring, this car was reduced to a steel sub-floor assembly by the simple means of taking a cutting torch to the body. Touring's artisans then fashioned the aluminum body to the style developed by Spring's team."

The cut-in doors, Constable noted, provided "almost incredible ease of entry into the low-slung car. The biggest man, complete with hat, simply drops into the car without any crouching, swivelling or stooping. Mr. Spring ardently wished this feature to go into the production car but for comparison purposes the left-hand side of the car was equipped with conventional doors.

"Ventilation [including ducted cooling vents for brakes] was another development well ahead of its time. . . .Similar in concept to the gull-wing Mercedes vent ducts [Spring owned and enjoyed a 300SL], those on the Hudson X-161 ventilate the car with the windows rolled up. The ducts also prevent compression effects due to slamming the doors with the windows closed." Cross-through ventilation finally came to Detroit ten or twelve years later. Notably, only the X-161 and the very forward-looking but similarly ill-fated Gaylord of the same period, had anything like it in 1954-55—and both were inspired by the gull-wing 300SL.

The X-161 used a higher rear-axle ratio than normal, hooked to a Dual-range Hydra-Matic and a Hornet Twin H-Power engine. It is not recorded what the X-161 weighed, but as it never ran the gamut of Engineering appraisal it probably was considerably lighter than the 3,600-pound production Hornet. If so it would have probably attained about 110 mph and provided 0-60 times in the ten second range, which would have been astounding in 1955. The 170 hp Hornet was already offering an impressive 21.1 pounds per horsepower. If the X-161 weighed, say, 400 pounds less (Spring claimed an advantage on this order for the original Italia) the figure would drop to 18.8. At 3,000 pounds it would be 17.6. The contemporary Jaguar XK-140, one of the fastest production cars in the world at that time, scored 15.8.

Also worth noting, the X-161 was very much like the production Nash, which became the basis of post-1954 Hudsons, themselves decorated with some Italia-like features. It had the same taut, rounded look as the attractive 1952-54 Nash cars, the same thin window frames and good visibility. It would have been fashionable alongside most 1955 production cars, and better looking than a lot of them. "Undoubtedly it would have been too much, too soon," says one former Hudson technician, "and never could have rescued the company, but I feel it would have sold better than the Jet did." Says Roy Chapin, "That X-161 Spring made from the Italia was an absolute knock-out, just like the Italia only with four doors. I got a laugh when Dick Teague [AMC vice president for Styling] proposed a four-door Javelin in 1970, for the ongoing Ambassador or Matador. I said, 'History's repeating itself.' [The X-161] is dated—but it's still a terrific exercise in automotive design. But again the problem with it was much the same with the other cars in the lineup. It was a very costly car to make and couldn't command the price you had to get for it, coupled with the fact that the decision was made to put nothing but a six-cylinder engine in it."

That same old bug-bear again.

A car of outstanding character and attractiveness usually survives in good numbers, and such is the case—happily—for the Italia. Perhaps eighty

Only contemporary photograph of the stillborn X-161, a 1955 prototype.

percent of them have been located. Even more fortunate for history's sake, both the No.1 'prototype' and the four-door X-161 live on, the former owned by Vic Racz, the latter by New York Hudson-Essex-Terraplane Club member Hal Denman. Both were saved originally by the timely intervention of Frank Spring, who learned that AMC was about to scrap them, and arranged to purchase them from the company. Spring took the X-161 with him to California, where he moved shortly after the Nash-Hudson merger. A few years after Spring's death his wife sold the car.

The X-161 spent many hard years before being tracked down and restored by H-E-T club member Elliott Myerson. He found the car in a bad state of repair. It had been quite an adventure finding it. When located, the X-161 was abandoned and derelict. It needed a tremendous amount of work, including the installation of a Hudson engine—a Cadillac V-8 had been installed. The aluminum body was dreadful, the result of children climbing all over it. Once the body was repaired, Myerson painted the X-161 an Italia-like cream, instead of its original dark green. Myerson sold the car to another H-E-T club member, Elwin Muzzey, who later sold the X-161 to its present owner, Hal Denman.

Both the Italia and the X-161 were anti-climactic—in a sense, the Hudsons that weren't. The times were against them. "The Italia," as Mrs. Spring said later, "was Frank Spring's swan song. I cannot help but smile as I remember Frank returning home one night saying, 'Well, today I showed Mr. Barit my new design for the Italia's instrument panel.' Mr. Barit's remark was, 'It needs more flash, more sex appeal!' Frank always wanted to make his designs clean and simple, without extraneous ornamentation, for practical visibility, etcetera. We both had the relief of a good quiet laugh."

CHAPTER 6

The Hash and the End

DURING EARLY JANUARY 1954, the public relations departments at Hudson and Nash-Kelvinator were working at fever pitch. Messengers shuttled back and forth between Jefferson Avenue and Plymouth Road, exchanging memos and long drafts. Endless details, statistics and historical data were checked for accuracy and run through the legal departments. "Meanwhile," said Nash's H. E. Hallas, "we met with Hudson public relations people, including their outside counsel, to discuss the nature of the story and the method of distribution. We received the copy of their story a few days later, and began dovetailing. . . .We felt the Hudson story was much too long as a news release, and had previously noted our legal counsel's caution with respect to over-detailing the arrangements. Some of the Hudson paragraphs, while well taken, seemed objectionable on this basis.

"Twenty-four hours before the anticipated release point, one hundred copies of the final story were duplimatted in our office and run off confidentially in the printing department; [Staffman] John Pichurski was stationed downtown, with releases ready for delivery upon signal, and three messengers standing by. The city was mapped out and the messengers dispatched in such a way that each of the three wire services would be reached at the same time. Pichursky timed his own departure to arrive at the *Wall Street Journal* at the same moment the messengers were hitting the wire services. Thus the broadtape would not scoop the AP or one of the others and vice versa.

"This turned out to be nip and tuck. We received our go-ahead at 4:10, January 14. Dow-Jones closes at 4:30. Allowing ten minutes for the messengers to arrive at the wires brought the ticker to within ten minutes of its deadline. So at about 4:16, we began feeding the story by phone to Stanley Penn, who put a sentence on the ticker at a time. We were also crowding the 4:00 o'clock afternoon deadlines and at 3:00 o'clock agreed that the *News* and *Times* could pick up the story at the Penobscot Building at the same time the messengers were dispatched."

The story was wired to papers in New York, Chicago and Grand Rapids, and mailed to 1,750 other papers as a courtesy—since they would get it sooner from the wires. Coverage of all major news outlets in Detroit, including radio and TV stations, was completed in nineteen minutes. Nash-Kelvinator and Hudson sales departments had been alerted a few days in advance, and were assisted in making field and employee announcements; all Detroit employees had been advised by letters posted in advance to their homes.

FLASH: IN AN IMPORTANT EXPANSION MOVE, DIRECTORS OF NASH-KELVINATOR AND HUDSON MOTOR TODAY APPROVED CONSOLIDATION OF THE TWO COMPANIES, SUBJECT TO STOCKHOLDER APPROVAL. THE COMPANY, TO BE NAMED AMERICAN MOTORS CORP. WOULD BE FOURTH LARGEST IN AUTOMOBILE FIELD. NASH, HUDSON AND KELVINATOR WILL BE SEPARATE DIVISIONS. DEALER AND SALES ORGANIZATIONS RETAIN SEPARATE IDENTITIES.

George Mason, Ed Barit and George Romney on merger day, April 22, 1954.

Mason reads the announcement. To his left, Barit looks away and Baits manages a smile; an era was passing.

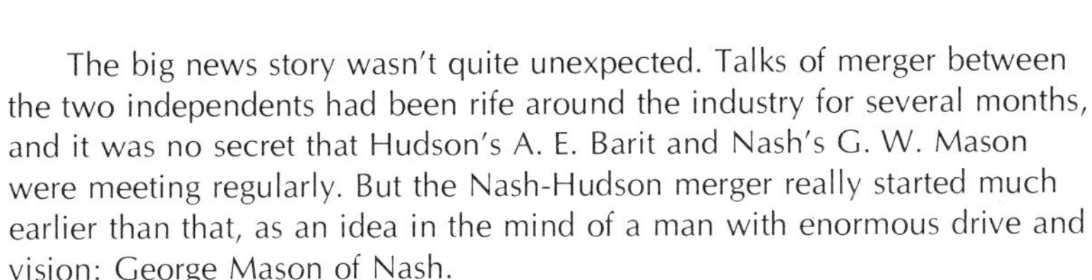

The big news story wasn't quite unexpected. Talks of merger between the two independents had been rife around the industry for several months, and it was no secret that Hudson's A. E. Barit and Nash's G. W. Mason were meeting regularly. But the Nash-Hudson merger really started much earlier than that, as an idea in the mind of a man with enormous drive and vision: George Mason of Nash.

In the romantic haze following World War II, and the dawn of the unprecedented postwar seller's market, Mason alone of the small manufacturers could see that the glut of business and bottomless reservoir of customers would not last forever. Rumors, unsubstantiated by this researcher, had Mason conferring with Ed Barit about an eventual combination of Nash and Hudson as early as 1946. However, one can definitely say the two met in 1948—Mason himself admitted as much in April 1954, just before American Motors became a formal entity:

"The idea was really born about six years ago, as a matter of fact, when Mr. Barit and I first talked about this proposition, although at that time, nothing came of it. About last June, Mr. Barit and I had many, many meetings." Then, laughingly: "He is tougher than hell. I have traded with [Charles W.] Nash for a good many years. I thought *he* was tough. But this guy Barit, he has a heart of stone! [Nevertheless] it has been a very pleasant negotiation. I think it is quite remarkable that we should arrive at a basis of exchange which would be acceptable to both shareholders." Mr. Mason, as was his wont, was being over-modest. His postwar scheme went far beyond the vision of a Nash-Hudson partnership. And had it not been for his sudden death just five months after the May 1 launch of American Motors, he might very well have brought it off.

To this writer, George Mason ranks as one of the most perceptive, intelligent and imaginative leaders of the postwar independents. He may have been the only one. Interestingly, *Newsweek* had this same impression in 1948, around the time Mason said he first approached Barit: "In some quarters of Detroit there is a suspicion that when the automobile industry returns to dog-eat-dog competition, big George Mason and his Nash will do all right. The principal reason is Mason himself. In between cigar smoking, trout fishing, duck hunting and gin rummy, the jovial 230-pounder has made a hobby of being a successful executive."

Big George was an appliance executive in 1937, when Charles W. Nash merged his company with Mason's Kelvinator (purportedly to get Mason's services, which he had failed to hire away). But Mason brought considerable automotive expertise with him. Born on a North Dakota farm in 1891, he was racing and selling cars as early as 1907. He graduated from the University of Michigan in 1913, serving at Studebaker, Dodge and Maxwell. At the latter he met Walter Percy Chrysler, and in 1924 he became general works manager for the brand new Chrysler Corporation. But the electric appliance field was just starting to mushroom then, and it soon diverted Mason's attentions. Joining Kelvinator in 1926, he became its president two years later when he was only thirty-seven years of age. Over the course of the next nine years he turned a faltering refrigerator maker from a million dollar annual loss to a million dollar annual profit. Aside from running the merged Nash-Kelvinator with consummate efficiency after 1937, Mason had served as president of the Automobile Manufacturers Association since 1946.

Roy D. Chapin, Jr., provided the author with a fond glimpse of George Mason. "He was a very warm, human kind of person," Chapin said, "and very direct—just the way he looks. He was also a very imaginative man, and a great outdoorsman. I knew him for years. In his duck marsh occasionally, when he was short one guide, he'd invite me down so I could set his friends' decoys for them and shoot their ducks for them and punt the boat, handle the dog and do all that kind of stuff. And I loved it. But he was interested in a lot of the same things I was—ducks, fishing and conservation. I'd known him as an individual long before we ever got in business together." Barit, at least across boardroom tables, was in Chapin's view "relatively distant. They were opposites, two completely different kinds of people."

The Nash Motor Company of Kenosha, Wisconsin was organized by former GM president Charles W. Nash in July 1916, when he acquired that city's Thomas B. Jeffery Company, manufacturer of the Rambler. From 1918 on, the cars used the Nash name. The early overhead valve sixes and fours were relatively unique in an industry populated by side-valve engines, and enjoyed good acceptance. Steady progress was recorded by the firm, which rose from fifteenth in the industry with 12,000 cars in 1917 to eighth place with nearly 21,000 by 1921. Nash absorbed Mitchell of Racine, and Lafayette of Milwaukee in 1924, and began to build some side-valve cars including the six-cylinder Ajax of 1925-26. Though these didn't do as well, the ohv Nash sixes in higher price ranges kept the firm well placed around tenth in the industry through the twenties. In the banner year of 1928, before the Wall Street crash sent the economy plunging, Nash built over 138,000 cars.

Though Nash was hard hit by the Depression—the bottom was reached in 1933 when 15,000 cars were built—it managed to hang on, and built some very impressive eight-cylinder cars later in the thirties. Helping its recovery was a low-priced 'new generation' Lafayette, reborn in 1934. That same year, Nash produced its millionth car. The products began to exhibit streamlining in 1935, with clean fastbacks, and skirted rear fenders. Before long Nash had added to its good looks mechanical marvels like coil-spring independent front suspension, seats that converted into beds, and an outstanding (Weather-Eye) heating and ventilating system—the latter was so good GM offered to sell Nash Hydra-Matics in exchange for its rights.

Mason's highly successful Nash 600 (for 600 miles to a tank of gas, a feat it routinely performed) introduced unitary construction in 1941 and sold extremely well with a list price of only $785. Nash built 80,000 cars that year, but the war put a stop to it and the company went into defense production. In 1946, the first full postwar year of car manufacture, it was back up to the 100,000 unit level, and production increased during 1947 and 1948. With the Farina-designed series of Airflyte models in 1949, Nash finally outproduced Hudson (which had usually led since the Terraplane days) and rocketed back into its tenth place industry position with 190,000 cars in 1950.

Like others in the industry, George Mason early saw the need for a small car. Unlike others, he built one that was not only good, but salable. His Rambler of 1950 was selling at a steady 50,000 units a year by 1952. In 1955, a vintage year for the industry, it would sell over 83,000.

Charles Nash died after the war, but even before Pearl Harbor he'd entrusted the running of Nash-Kelvinator to his competent associate and successor. With the war behind him, Mason hired his own heir apparent: George W. Romney, his assistant in the AMA. To join Nash, Romney had turned down an offer from Alvan Macauley of Packard, who promised more money but, in Romney's view, less potential. With the succession secure, George Mason now began to look ahead.

"There was a time prior to the merger, right after the war, when everybody figured they'd get rich in a hurry because there was such pent-up demand," Roy Chapin says. "There was a point somewhere in there that Mason approached Barit, but Hudson was planning the 1948 product and one can't blame Barit for saying no at the time. In retrospect, Mason did have a grand scheme in which Hudson, Nash and Packard were all going to be together, and that's the only thing I know of. It was like most of those schemes—you never could get all three people in the same frame of mind at the same time."

But there is more to the merger story than that. It involved not only the Detroit companies but another independent in Indiana—the largest producer of automobiles outside of the 'Big Three.' The fact is that Mason envisioned something bigger than the Detroit amalgam—he wanted to pull in Studebaker as well!

The writer learned this remarkable story quite by accident, while interviewing former Packard president James J. Nance. It was common knowledge, of course, that merger talks had occurred between Packard and Nash, or American Motors, and some sharing of components actually did take place though no formal alliance was ever consummated. But the Nance conversation seems to indicate that more than Packard was involved.

Alvan Macauley and Hugh Ferry, chairman and president, respectively, of Packard, had been looking for a new leader almost from V-J Day. After Romney turned them down, they cast their eyes on the dynamic president of Hotpoint, James J. Nance*, who had turned that company into one of the sales leaders in the appliance field. "Hugh Ferry arranged a meeting in Chicago to talk to me through the president of the Harris Trust Bank, who was a friend of George Mason's," Nance states. "I was told then that George Mason had the idea of putting the four independents—Hudson, Nash, Packard and Studebaker—together. Their approach to me was a first step of putting this together as a single multiple-line producer in the postwar period. Mason had been trying to do it for a year or two [Nance met with Ferry in 1951] and had not been able to negotiate a deal with Harold Vance of Studebaker.

"The offer to me was to put Studebaker and Packard together so that later on the four companies could be brought together as a postwar multiple-line automotive manufacturer. Up to that time my knowledge of the automobile business had been limited to General Motors. I grew up with Frigidaire which at that time was just an island in the automotive world.

"George Mason had the original idea and never lost interest in doing it. He had picked up Hudson which was then for sale and being shopped, that was no secret. Everybody in the automobile industry knew [Holland's] Queen Wilhelmina owned eleven percent of Hudson and wanted to sell.

"We agreed that Mason would take the Hudson-Nash end, and I would put Studebaker-Packard together, then we'd fold the two pieces together into one company. The name was to be American Motors. *I wouldn't have gone into it just to take over Packard.*" (Italics are the author's.)

Why did Mason ask James Nance to be copartner in the grand scheme? Well, he obviously knew Nance and appreciated his abilities—they both had been in the appliance business. He had personally broken the ice with Hudson, in 1948. And he needed an intermediary to fold-in Studebaker.

*Mr. Nance's comments are used with permission, from an interview copyright J. J. Nance, 1976.

Unit bodies were again the rule for '55, but now they used the Nash shell which Hudson panned only a few years before.

In October 1954, *Business Week* printed this shot in the dark: "There's no doubt that Mason, one of the auto industry's early pioneers, was the strong man at American Motors. He instigated the Nash-Hudson merger that brought the company into being . . .Detroit feels that his powerful personality was a deterrent to any merger with Studebaker-Packard, which is also dominated by powerful personalities. Mason and Studebaker-Packard's top managers might never have reached agreement over who should boss the new company."

This was not exactly right, but it was about as close as anyone in the press came. It wasn't Studebaker-Packard that had Mason stymied, it was *Studebaker itself*—before the Packard connection. Those strong, powerful personalities were Mason on one hand and Studebaker's Paul Hoffman (chairman) and Harold Vance (president) on the other. James Nance was the ideal middle man. A brilliant innovator and champion salesman with no auto industry skeletons in his closet, Nance could probably bring

The 1955 Hudsons benefited from Nash's Weather-Eye heater-ventilator and used an individual dashboard similar to that of Jefferson Avenue's '54 models.

Studebaker into line. And he did: Packard officially purchased Studebaker in November 1954. All was going according to plan. But on October 8, 1954, George Mason died.

"If George Mason had lived, I believe we'd have put them together," James Nance told this writer. "But . . .Mason died very suddenly of pneumonia. He was succeeded by George Romney, and there were no further discussions of a merger. He did propose a 'love without marriage' arrangement that would have called for consolidation of many functions of the two companies, but not a merger. It was the conclusion of our management group that the plan was not acceptable or workable because of the difficulties of administration, and negotiation required on practically every item, so Romney went his own way. Romney took the small car they had, the Rambler, and did a superb job of promoting it. This was the car Mason was going to drop [This is disputed by several AMC executives. George Romney would not comment to the writer.] when the final merger was made because it was duplicating the Studebaker, which was to be the entry into the small car field to compete against Chevrolet and Ford. There is reason to believe that if George Mason had lived, the plan would have been consummated."

It has been reported in at least one Hudson article that during the merger negotiations, Packard objected to Hudson's presence as a partner. This is not recalled either by Chapin or Nance. "I certainly had no objection," says the latter. "As a matter of fact, the job was divided up into phases. Since Mason did not negotiate with Studebaker, the first step was to put Studebaker and Packard together. That was my job. We chose Packard as the surviving company because Packard had some huge defense contracts building jet engines which we figured could carry us through the period where we'd get consolidated and get refinancing."

Wouldn't this huge combine have created product duplication? "We didn't think so," James Nance states. "We would have had four lines of cars. Packard would have been moved up to the top end of the line: Studebaker to the bottom end of the line; Hudson and Nash in the second and third positions. You've got to remember in those days, up until 1956 or 1957, there were very definite price classes in the automotive industry. Historically a family man started out with a Chevrolet, moved to Pontiac, moved to Oldsmobile and Buick and then to Cadillac. General Motors for years had the slogan, 'A car for every purse and purpose.' This was the multiple line pattern that Ford was following when they brought out the Edsel, and started to make separate kinds of cars out of the Continental and the Lincoln. So we had agreed on the four lines of cars and two dealer networks." Mr. Nance emphasizes that there was no thought of all dealers handling all four makes, because agencies geared to low-priced lines could not sell the luxury models. This follows the GM pattern, where Cadillac dealers sometimes sell Oldsmobiles and Buicks, but rarely Chevrolets and Pontiacs.

Evidently these plans were on a very high level, and discussed in hushed confidence, because nobody except Mason and Nance had realized Studebaker was a potential fourth partner. Roy Chapin, when questioned whether Studebaker was mentioned, says "not to my knowledge." But he also points out that he was only assistant sales manager at Hudson, not privy to topmost dealings. He does recall "one concept under which Nash was going to make the small car because Mason was small-car oriented, then Hudson would make the medium-priced car and Packard the expensive car. That was the concept. Packard was going to make the V-8 engines, Hudson the six-cylinder engines, Nash the small six, that was about the way it was. What the corporate structure was I don't know."

This doesn't coincide with Nance's impression that the four-company merger would eliminate the Rambler, but the time frame may have been different. The possibility of *not* bringing Studebaker in had to be one of Mason's contingencies. Surely Nash had a very salable small car in the Rambler. If Studebaker came in, fine, the Rambler would disappear or become a luxury compact. If not, well, the Rambler would carry on. It makes sense.

The whole Mason idea makes sense, really. Consider this comment from the contemporary press: "Neither Hudson nor Nash offers a car in the high-price bracket. Neither has a truck. And neither has more than the beginnings of a network of wholly owned suppliers. [But] Packard has the high-priced line and Studebaker the truck, plus the low-priced Champion—which, over the long pull, might be even a better competitor of Chevrolet, Ford and Plymouth than the Rambler. Independent suppliers abound in the Michigan and Ohio area . . ."

Maurice D. Hendry quotes one source as saying that during the discussions "a certain individual . . .convinced Packard that they should be the big wheels in the new company, as they had the V-8 engine, which was all that Nash and Hudson were really after." The Packard engine was still a year from production at that time, though it was closer than any Nash or Hudson V-8.

"My impression on the V-8 was that it was what both of us were looking for," says Chapin. But the author tends to doubt the rest of Hendry's source's statement. It doesn't jibe with the character of George Mason—nor with this writer's experience with James Nance, who does not appear to be the type to give up the presidency of Hotpoint simply to preside over the decline of an unmerged Packard. Besides, there doesn't appear to be any supportive evidence for this.

Mr. Mason *was* disturbed, several insiders state, when reciprocity plans with Packard for the exchange of components became a one-way street: Packard willingly sold AMC its V-8 engine in 1955, but refused to contract for anything in exchange. This may have fueled the scuttlebutt around Detroit after Mason's death that an AMC-SP tie-up was more, rather than less, possible. But George Romney, who succeeded Mason on October 12, 1954, flatly denied any plans for further mergers. "There are no merger discussions, and there have been none with Studebaker-Packard [he didn't say Packard before Studebaker]. We are highly diversified . . .we know we have a bright and prosperous future as American Motors. It's not based on merger with any other company." Some publications, like *Newsweek* and *Business Week*, publicly didn't believe him, but as time proved, George Romney spoke the truth.

And so Hudson would go its way with Nash. Though it might be powered by a Packard engine, it assuredly would not be sold by Packard dealers—or vice versa. The 1955 Hudson, win or lose, would be a product of American Motors.

Even the Nash-Hudson merger took a long time to work out—six months, in fact, from the start of negotiations. Barit was indeed tough, though as Roy Chapin says, "He was a realist; he knew you couldn't run the company with the kind of volume we were projecting with two different cars." But Barit dearly loved the Jet (so did first vice president Baits), and they went round and round with George Mason over it.

Mason's attitude that the Jet absolutely *had* to go was entirely justified. The Rambler was clearly more marketable—the Jet wasn't even selling well enough to amortize its tooling costs. Barit offered to settle for the inclusion of Jet investment in Hudson's assets, but Mason insisted it be killed and the investment in it liquidated before Nash would go further toward a combined operation.

In the end Barit had to give in. He simply had too few cards to play. "Hudson, on paper, was in a poor bargaining position," *Business Week* explained. "Its annual report for 1953 hadn't been issued, but for the nine months ending last September 30 the company lost $838,100 after a tax credit of $760,352. Sales continued to decline in the last quarter of 1953. . . .Further evidence of Hudson's position came when it disclosed to the New York Stock Exchange that it had entered into a $20 million line of credit, expiring June 30, 1955, and had already used $11.6 million of it." To pay off a Nash debt of $40 million and a Hudson debt of $29.6 million, American Motors drew down $69.6 million of its own $73 million line of credit. That didn't leave much.

Under terms of the agreement, three shares of Hudson stock would convert into two shares of American Motors, while each share of Nash-Kelvinator would continue as one share of AMC. Nash, Hudson and Kelvinator would operate as seperate divisions of American Motors; dealer and sales organizations would retain separate identities. Ed Barit would serve as a director and consultant, with Mason president and board chairman.

As of September 30, 1953, Hudson and Nash-Kelvinator together had assets of over $357 million and working capital of over $100 million (two-thirds of it Nash's). Combined sales of the period from October 1, 1952 to September 30, 1953 exceeded $680 million, with total production of 244,507 vehicles. Kelvinator produced over 638,000 electric household and commercial appliances in the same period.

Mason and Barit made a joint announcement: "Both of us see as the result of this move greater opportunities for Nash, Hudson and Kelvinator dealers. . . .Our 16 plants will include three large automobile body plants, and complete engine manufacturing facilities and automobile assembly plants. It will have foundry and forge facilities, and will be equipped to build a large share of other major parts. The combination will be one of the most highly integrated in the automobile field. Among the advantages of the move are pooling of executive abilities, research and engineering resources, and purchasing power; opportunities for new manufacturing economies and improved methods; reduction of the individual overhead and administrative charges; a diversification of products which tends to stabilize earnings in periods of high and low business activities, and a spreading of tooling costs over more units with less cost per unit.

"We are proud to contemplate the association of three such fine names as Hudson, Nash and Kelvinator in a single American enterprise. The tradition for which these names stand and a vast storehouse of know-how which the two companies have accumulated in their nearly half-century of successful operations have been a source of mutual respect for many years and of mutual enthusiasm during our negotiations."

Both Mason and Barit said their respective boards were unanimous in recommending the merger. Hudson stockholders weren't. Well, not entirely. In their February 1954 poll, 245,000 shares voted no against 1.4 million in favor and 300,000 uncommitted. Hudson announced that dissenters could, if they liked, sell out at $9.62½ per share—and a New York attorney with 2,000 shares promptly demanded $30 per share. A stockholder suit asked the courts to block the merger on the grounds that Hudson's financial statements, issued with proxy material, were "misleading and failed to give the stockholders complete information concerning the company's operations and financial position." But these protests were disallowed, and in the end only seven percent of Hudson's stockholders elected to cash in their shares.

Hudson people really had something to deplore when Jefferson Avenue management pulled a surprise in May. In a circular letter, carefully routed "TO HUDSON SUPERVISION" (only), management advised that "utmost economies . . .are necessary. While the Hudson and Nash cars will be kept as distinct lines, changes planned for the 1955 models of both lines of cars necessitate the centralization of body production and trim and car assembly at the Milwaukee and Kenosha plants of American Motors. Hudson engines and other components will continue to be built in Detroit, and shipped to Kenosha.

"Every possible effort will be made to cushion the necessary readjustment. Lay-off of employees will be made successively as various parts of the 1954 model program are completed . . .a placement service is being established to contact and seek the cooperation of other companies in the greater Detroit area in finding jobs for Hudson employees whose services no longer are required."

It was a very nice letter considering the bleak news—except for its timing. "Because of our years of association," it concluded, "we felt you are entitled now to a clear statement from us." Sure they were. That letter was distributed on May 27. Twenty-four hours later, the same message was made public. With this 'advance warning' Hudson repaid supervisory personnel for years of service. As for the line workers, they had to read about it in their morning papers.

On October 29, 1954 the last Hudson—appropriately a Hornet—rolled off the line at Jefferson Avenue, forty-five years almost to the day since the first one had come off. Management refused to disclose the ultimate number of lay-offs—Hudson engines and some defense work would continue at the plant—but the *Detroit Free Press* reported that Jefferson's 9,000 workers would be down to 3,000-4,000 by the end of the year. Several hundred transferred to Kenosha; a thousand found work at Chrysler and Ford. The rest became unemployed. Inevitable, perhaps—the expensive jigs at Jefferson, unsuited to integrated body production and impossible to transfer economically, left little alternative. But the announcement had been handled in a shoddy way by what was left of Hudson management.

The people who count beans must have been heartened by the concentration of body production in Kenosha—the merger would have offered few cost savings otherwise—and some analysts now wondered if American Motors could standardize on engines. That would take a little time yet. But they were working on it. They were also working on the new 1955 Hornet and Wasp, which would certainly wake up the traditional Hudson buyers. Ironically, the restyle that Hudson had never been able to accomplish was finally being done for them, by those other unit body people whose products they had always criticized as too light, too high and too clumsy. It wasn't, therefore, very hard to recognize the new Hudson for 1955. It was a Nash. Or as some called it, a Hash.

Model Body Style (passengers)	Price	Weight (lb)
Series 54 Metropolitan		
hardtop (3)	$1,445	1,843
convertible (3)	1,469	1,803
Series 55 Rambler Deluxe		
sedan, 4dr (6)	1,695	2,567
club sedan (5)	1,585	2,432
Suburban wagon (5)	1,771	2,528
Series 55-1 Rambler Super		
sedan, 4dr (6)	1,798	2,570
club sedan (5)	1,683	2,450
Suburban wagon (5)	1,869	2,532
Series 551-2 Rambler Custom		
sedan, 4dr (6)	1,989	2,606
Country Club (5)	2,098	2,685
Cross Country wagon (6)	1,995	2,518

Series 3554-1 Wasp Super		
sedan, 4dr (6)	2,290	3,254
Series 3554-2 Wasp Custom		
sedan, 4dr (6)	2,460	3,347
Hollywood hardtop (6)	2,570	3,362
Series 3556-1 Hornet Six Super		
sedan, 4dr (6)	2,565	3,495
Series 3556-2 Hornet Six Custom		
sedan, 4dr (6)	2,760	3,562
Hollywood hardtop (6)	2,880	3,587
Series 3558-1 Hornet Super		
sedan, 4dr (6)	2,825	3,806
Series 3558-2 Hornet Custom		
sedan, 4dr (6)	3,015	3,846
Hollywood hardtop (6)	3,145	3,878

Hudson Rambler Country Club hardtop.

Hudson Rambler two-door sedan, 1955.

The 1955 Hudson Rambler Custom sedan.

To sort out this incredible mélange, it can first be said that the top four entries were all badge-engineered Ramblers. Nash's products not being the subject of this book, we will not deal with them in detail here. The Hudson versions were identical to the Nash right down to the exact dollar price. The Metropolitan, introduced in March 1954 on an eighty-five-inch wheelbase, was powered by a forty-two horsepower four-cylinder Austin and built for Nash in England by the body assemblers Fisher & Ludlow. The Ramblers, updated from the year before with new oblong grilles and egg-crate meshing, were on one hundred-inch wheelbases and ran an L-head six of 195½ cubic inches (3⅛×4¼) producing ninety horsepower at 2800 rpm. Overdrive and Hydra-Matic were optional on Ramblers at $104 and $179 respectively, and the hottest-selling model was the Custom Cross Country, a neat little wagon that was rapidly growing in popularity. Not all Nash dealers were happy to see the former competitor down the street selling their cars with Hudson badges and hubcaps. Hadn't the company told them that Hudson and Nash would remain 'separate?' There were a few cases of roiled waters because of this.

American Motors provides no records of how many Metropolitans bore Hudson badges and how many Nash, but some exact production figures have come to light on the Ramblers. Of 81,237 1955 Ramblers (including export models and thirty-five delivery vans), 25,214 were labeled Hudson

The 1955 Hudson Rambler Custom Cross Country wagon.

The 1955 Hudson Wasp sedan.

and 56,023 Nash, an approximate two to one majority in favor of the latter. Metropolitans probably had a similar breakdown, but some of the 1954's were sold as 1955's after November 26, 1954, which complicates calculations. A complete breakdown of the 25,214 Hudson Ramblers will be found in the Appendix.

"New high style, new V-8 power" was the way the big Hudsons were described. With rather attractive styling on the basic Nash body, they retained the familiar Hudson dashboard layout, which was a clever reference point to their predecessors and probably used up some extra parts. Together with the Mets and Ramblers the new Hudson line was the largest and broadest in price ever, and easily represented "the most sweeping model changeover in Hudson's 46-year history," as N. K. VanDerZee put it. The 202-cubic-inch L-head six—all that remained of the ill-fated Jet—powered the Wasp models. The familiar 308-cubic-inch six was used in the six-cylinder Hornet, and a Packard-supplied 320-cubic-inch V-8 (3 13/16×3½ inches bore and stroke, 7.8:1 compression, 208 hp at 4200 rpm) powered the Hornet V-8. With Twin H-Power available on the six, Hornet buyers had three types of engines to choose from for the first time in history, the two sixes measuring, as before, 160 and 170 horsepower. The Jet-engined Wasp, interestingly, was rated at 110, Twin H-Power bringing it up to 120.

Both cars featured Nash's fully wrapped windshields (the largest in the industry at the time and notably free of distortion), flanked by fairly narrow, vertical corner posts and a broad chrome molding that formed a 'visor' at the top. Front fenders were higher than the hoodline—a feature pioneered by Packard in 1951 and almost made part of the 1955 GM line—and were cut away around the wheels "to provide a short turning radius." An interesting point, since the comparable Nashes continued to hide most of their front wheels; but this was an inexpensive die change that helped provide good visual identification between the two makes.

"The new Hornets and Wasps have been given a smart and functional rear quarter treatment," the press releases bubbled. "The broad counterbalanced trunk deck covers a roomy luggage compartment. The large taillights are mounted high and to the extreme rear of the fenders for excellent visibility. Luxurious interiors of smart Bedford cords, genuine leather and vinyl are complimented by an array of eleven solid exterior colors and eleven two-tone combinations.

"The Hornet and Wasp series will be offered in four models, including four-door family sedans and custom hardtops. The wheelbase of the Hornet is 121¼ inches; overall length 209¼ inches; height 62¼ inches [a couple inches *higher* than the 1954 Hornet!]; and width, 78 inches. The Wasp's wheelbase is 114¼ inches; overall length 202¼ inches; width 78 inches; and height, 61¾ inches."

The concurrent Nash Statesman, Ambassador six and V-8 models followed the same Super-Custom body lineup as the Hudsons but were powered by overhead valve Nash six-cylinder engines, ranging from 110 to 140 hp, and the shared V-8. The latter was about $50 cheaper in the comparable Nash than in the Hudson. Nash sixes were also somewhat less expensive, by up to $75. One Hudson feature that fans were happy to see

Hudson Hornet Hollywood hardtop, 1955.

retained was the Triple-Safe braking system with dual hydraulics and mechanical back-up.

Motor Trend had some even-handed comments on the new Hudsons, but though the cars were still advertised as "the stock-car champions" the magazine made it clear they weren't the Hudsons of old. "Company officials make no bones about all the bigger models of Hudson and Nash using the same basic body structure, and proudly point to the resulting savings in manufacturing costs. There wasn't much sense in continuing with Hudson's old Step-down design, as that had been vamped and revamped since 1948. The short, narrow Hudson Jet was a styling fiasco [one notices how the monthly slicks always speak their mind *after* the car is out of print!] and hence an unpopular model, despite its outstanding performance, so it, too, was dropped. Nash's current Farina design is just turning four years of age, but with the radical changes this year, it is hardly fair to call it the same design. The amalgamation of an 1,100-square-inch wrap-around windshield is little short of an engineering miracle, especially considering the fact that the project was started only a few days before George Mason's untimely death on October 8, 1954. In fact, it was one of Mr. Mason's last orders.

"The Hudson has the more orthodox front end design [Nash had shoved its headlamps inboard, which amazed people], but it is equally pleasing. . . .The Hudson dash uses last year's very legible speedometer mounted high in front of the driver. When a radio is installed, both [Nash and Hudson] have twin front speakers."

The big question among Hudson lovers wasn't the restyle—it was whether the Hash would provide the same peerless roadability as the Step-down Hudsons. *Motor Trend*'s verdict was a blend of fact and press release: " . . .whatever has been lost in the handling department (few except Marshall Teague would notice it) has been more than made up by the car's new-found liveability." The magazine went on to praise the Hash's improved ease of entry and exit, Nash-style seats that reclined into beds, better visibility over the Step-down ("which caused some to suffer claustrophobia"). The big V-8 engine, naturally, provided better acceleration than the 1954 Twin-H Hornet, clocking 0-60 in twelve seconds and the quarter mile in 18.5 at 74 mph. This was not among the fastest of '55 test cars—Packard's Ultramatic, providing clutchless shifting for the big AMC cars, was not outstanding in this respect—but the Hornet did shift about as smoothly as any car. *But how was the roadability*? Aha . . .

"Old Hornet's reputation for 'stickability' is carried on by new model to surprising degree. Except for tendency to wander away from crown of road, steering is unperturbed by surface variations or wind gusts. Cornering is positive with very little lean. Firm springing smooths out dips in old Hornet tradition. Car, heavily loaded during travel phase of test, bottomed only on a pot-holed stretch of U.S. 40 in Utah that would have dismayed a Jeep."

Motor Trend summarized: "We talked to a number of loyal Hudsonites who traded in their 'Step-down' cars on the new one. They seem pretty well satisfied as a whole, pointing out that they now have a great increase in 'liveability' without sacrificing much roadability." Hudsons for '55 wouldn't run in NASCAR races anymore, now that that 'little extra' in the handling

Hudson Metropolitans, the ultimate example of badge engineering.

department was gone. "This should be balanced against pleasures of owning a cleverly engineered car whose duplicate you won't meet going around every corner." (You wouldn't meet Marsh Teague in one, regardless. In 1955, at a NASCAR circuit, he took out a new Hornet, turned one lap at speed, drove it into the pit and walked away, never to enter one again.)

Overall, the Hudson sales trend in 1955 was up. About 32,000 Hornets, Wasps and Jets had been sold in 1954; the figure for the 1955 models was just 20,321; the Hudson Ramblers, however, did much better. With 25,214 of those, the marque could claim 45,535 model year sales—a healthy fifty percent improvement. The Jet, it is necessary to mention, never did reach 25,000 in a model year.

The first Hash has probably been overly condemned by the overly rabid. It was a good car, soundly engineered, offering the highest development of an L-head engine in the Twin H-Power Hornet six, reasonably styled for the period and fitted with a host of fine Nash and Hudson equipment like Weather-Eye heater/ventilator, Triple-Safe brakes, seat-beds and deluxe long-haul touring accessories. It was nicely finished for the most part, with luxurious interiors in attractive colors. One is reminded of a favorite phrase uttered around the same time by Edgar Kaiser; "Slap a Buick nameplate on it, and it'd sell like hotcakes."

The 1956 Hudson was another matter.

Model Body Style (passengers)	Price	Weight (lb)
Series 56 Rambler Deluxe		
sedan, 4dr (6)	$1,672	2,891
Series 56-1 Rambler Super		
sedan, 4dr (6)	1,773	2,906
Cross Country wagon (6)	2,047	2,992
Series 56-2 Rambler Custom		
sedan, 4dr (6)	1,884	2,929
Country Club wagon (6)	2,289	3,095
Cross Country wagon (6)	2,136	3,110
hardtop, 2dr (6)	2,038	2,990
Series 3564 Wasp		
sedan, 4dr (6)	2,214	3,264
Series 3565 Hornet Special		
sedan, 4dr (6)	2,405	3,467
Hollywood hardtop (6)	2,512	3,488
Series 3566 Hornet Six		
Super sedan, 4dr (6)	2,544	3,545
sedan, 4dr (6)	2,777	3,636
Hollywood hardop (6)	2,886	3,646
Series 3568 Hornet V-8		
sedan, 4dr (6)	3,026	3,826
Hollywood hardtop (6)	3,159	3,026

The extra line in the form of the new Hornet Special did not indicate a decision to increase Hudson market penetration—anything but. What it stood for was a new V-8 of 190 horsepower—engineered and built by American Motors in Kenosha—which replaced the Packard V-8 in mid-season. This act had the approval of everyone at AMC—Packard and Studebaker-Packard hadn't bought much in return since the 'reciprocity' plans had been worked out except a few minor bits and pieces.

But the styling of the 1956 Hudson was a horrendous attempt to rescue the old triangular motif of earlier years. Called the 'V-line' it was reportedly the work of Edmund E. Anderson at AMC styling, and the best that can be said of it is that, well, some other '56's were worse. But not many. The

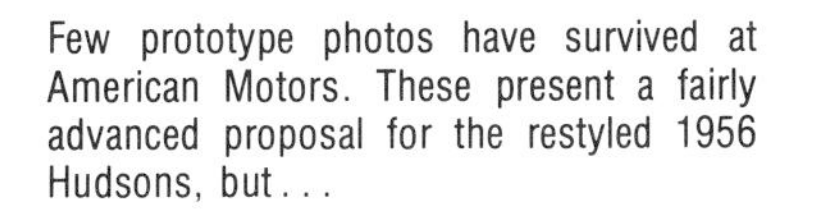
Few prototype photos have survived at American Motors. These present a fairly advanced proposal for the restyled 1956 Hudsons, but . . .

. . .when the new face did appear, it was even uglier.

rework involved a huge die cast, egg-crate grille, interrupted by a V housing the Hudson emblem at the top and dipping in another V-shape at the bottom, with side-visible parking lights thrusting inward through huge chrome housings. The hood ornament was a bizarre pointed affair, and dummy air scoops were grafted onto the chrome-capped front fenders. On the sides, the V motif was repeated in the two-tone division line, and a gold anodized aluminum panel ran back from the rear door. At the rear, the taillights were similarly glued on, complete with little chrome fins, a pitiful reply to the growing craze for this styling gimmick, and designed neither to out-look nor out-measure the competition. Inside, the 1954 instrument cluster was replaced by one built into the dash, not the same as the comparable Nashes which used a dash similar to 1955. Mechanically the Wasp Six was up to 120 hp and 130 hp with Twin H-Power, the Hornet 165 and 175 hp, ditto.

The 1956 Wasp sedan.

The 1956 Hornet Custom sedan.

On the road, the 1956 Hudson seemed to roll more than the 1955 and was anything but the handler the Step-down had been. The new engine, however, was a good one: 250 cubic inches, 8:1 compression, 240 foot-pounds of torque. "The engine looks quite conventional," stated one report. "Its short, fairly wide rocker arm covers follow the 90-degree cylinder angle. Mounted on a four-point system (two brackets fore and aft), it has aluminum-alloy, steel-insert three-ring pistons, a five-bearing crankshaft, cast iron heads. Present form is limited to the engine as described here; there's no power-pack option. With a displacement increase, carburetion could be vastly improved, and compression ratios could skyrocket. This smallest of U.S. V-8s would be a natural in the Rambler at this displacement, but will have to grow to meet sales demands in the present field."

Interestingly, this engine or something like it might have ended up in a Kaiser several years earlier, had that company the money to finance it. It was the design of Dave Potter, who had worked with engineer Harold Bullard on a Kaiser-Frazer 288-cubic-inch V-8 in the early fifties. "Some of the design ideas of the Kaiser 288 did become the industry design some years later," Potter told this writer. "Whether this was carried by Kaiser engineers to other companies when Kaiser closed or was just the natural progress of the best way, is hard to say. I can, however, think of some design ideas I used in creating the AMC V-8 that are found in all modern V-8's: the use of green sand casting for the tappet chamber and front face, instead of baked oil sand core, to achieve greater dimensional accuracy as well as low foundry cost, for example. Or using the main oil gallery drilling to intersect by half its diameter the tappet holes on each bank, to provide metered oil for the hydraulic tappets."

Most reporters found the new AMC V-8 unequal to the performance of the big Nash and Hudson, and nowhere near the former Packard-built unit. Their 0-60 figure came to 14.6 seconds (slower than the Twin-H Hornet!) and the quarter mile was clocked in 19.7 seconds at 69.7 mph. This Hornet Special—unlike that Hornet Special of 1954—was a pig.

For 1957, Ramblers were listed as a separate make and the Hudson line shrank to V-8 Hornets exclusively. But the latter was now up to 327 cubic inches, belting out 255 horsepower and 345 foot-pounds of torque, so performance was much improved and back up to 1955 standards. Hydra-Matic, which had replaced Ultramatic on the V-8's when AMC dropped the Packard engine, was optional as was overdrive. Three-speed synchromesh remained standard. The 1957 Hudsons continued to offer Triple-Safe brakes, and now added air conditioning to its list of accessories that already included power brakes, steering and windows.

Model Body Style (passengers)	Price	Weight (lb)
Series 357-1 Hornet Super		
sedan, 4dr (6)	$2,821	3,631
Hollywood hardtop (6)	2,911	3,655
Series 357-2 Hornet Custom		
sedan, 4dr (6)	3,011	3,678
Hollywood hardtop (6)	3,101	3,693

The Hornet Hollywood hardtop, 1956.

Styling on the '57's was improved over the '56's, though the anodized trim (silver this time) was still found on Customs. A V medallion was added to the grille, as if it needed more decoration—the real reason was to emphasize the V-8 engine that the public was by now literally demanding. The scoops over the headlamps were gone, replaced by dual, reversed chrome fins; at the rear, tacked-on fins were painted the body color and given more of a taper, which better integrated them with the car's lines. The Custom Hollywood introduced, with the comparable Custom Ambassador, three-tone colors for the first time on Hudson or Nash, and new 'Tri-Tone-interiors in metallic-weave nylon, vinyl or leather were color-keyed to match the exteriors. One of the more attractive three-tones was a combination of metallic avocado, black and Mojave yellow (a light cream); but if the buyer didn't like that, Hudson offered a total of thirty-two one-two- and three-tone combinations to choose from. The cars were still very luxurious inside, with sixteen-inch-wide center armrests in the rear, Nash's pullout glovebox, right and left ashtrays as well as radio speakers, and dual safety-padded sun visors. As was typical in cars of the age, the less costly Super models were more cleanly styled: available in two-tones or one-tones only, they lacked the garish silver side trim—on two-toned models—this area was replaced by the roof color.

During model year 1956, 20,496 Hudson Ramblers were produced, along with 10,671 Hornets and Wasps. For 1957 the figures were: 114,084 Ramblers and 1,345 Hornets. (Breakdowns are in the Appendix.) A notable price cut between 1956 and 1957, and the tapering off of model choices, led observers to wonder whether there would be a 1958 Hudson. The industry expected one as late as September 1957. Reports indicated that the '58 Hudson would use a larger displacement AMC V-8 with over 300 hp on a slightly shorter wheelbase. Air suspension was to be introduced, according to several reports. Space utilization would be improved, and both Nash and Hudson would be very competitively priced. Style-wise the cars would standardize dual headlights, and offer a more distinctive and cleaner front end. Four-door sedans would predominate, but a station wagon and a two-door sedan were to be offered.

Artists' impressions of the 1958 Hudson carried the familiar arrow-shaped front fender motif but no tail fins. Drawings indicated more glass area and a full-width, squared-off, egg crate-type grille. When the 1958's did come out, the origins of these sketches became obvious: They depicted the general lines of the 1958 *Rambler*.

The names Nash and Hudson were nowhere to be seen, but most of the advance rumors had been confirmed. Wheelbases were a few inches shorter (1957 Hudson was 121¼ inches, 1958 Rambler Ambassador 117 inches) and overall length was little changed. Air suspension was not an option, but dual headlights were standard and the grille, cleaner and distinctive. A trace of V-Line Styling remained in the lower Ambassador grille and Rambler Custom side trim. But of the famous white triangle, and the name that had graced some of the finest American cars for nearly half a century, there was no sign. Hudson, the marque, was gone for good.

According to several sources at AMC there was a lot of soul searching about the decision to drop both Hudson and Nash as makes. Granted that the future seemed to lie in small cars, which the competition wasn't selling, but couldn't the Ramblers be given Hudson and Nash designations?

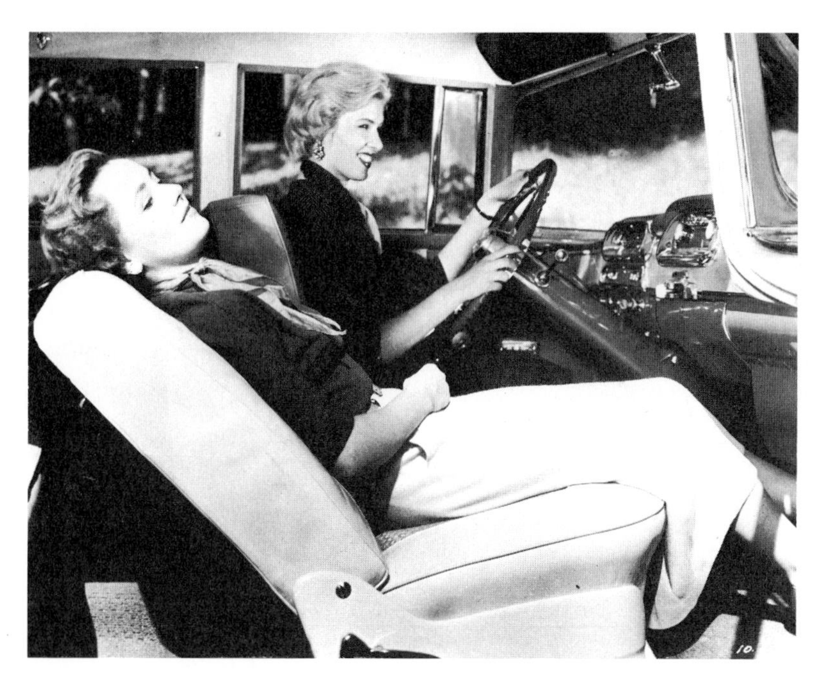

Teaming up with Nash also brought 'airliner reclining seats' to Hudsons, offered as standard on all Customs; optional on Supers. The seats had five different positions and air mattresses could be bought as an optional extra.

The 1957 Hornet sedan.

Conservatives naturally based their argument on history: Together Nash and Hudson had ninety years of production behind them, and many great accomplishments. Those favoring change replied that Rambler, not Nash or Hudson, had become the name people related to and the car that they were asking for in rapidly increasing numbers. Rambler, after all, had been around even longer than Nash and Hudson, although it had been absent for a long interregnum between 1914 and 1950.

In the end, management decided in favor of Rambler, retained Ambassador as a designation for the best model thereof, and dropped both Hornet and Wasp as well as Statesman. And, of course, Nash and Hudson.

In deserting the big cars and staking AMC's future on the Rambler, George Romney was taking a calculated risk. He was certain that a steady market for smaller, U.S.-built cars existed, even luxury versions thereof. Rambler sales provided him with convincing evidence. Better still, the Big Three didn't have anything at all in this field. "Frankly," said Romney in announcing the demise of Nash and Hudson, "we very nearly got knocked out of the automobile business in the process of succeeding in our audacious decision to challenge the Big Three by flanking their production position." AMC had had two hard years, and the tough decision probably was inevitable. In the fiscal year ending September 30, 1955, the company lost close to $7 million on sales of $441 million; in the next twelve months it lost close to $20 million on sales that were down to $408 million.

A. Edward Barit resigned his directorship in 1956 and retired. "He was terribly disappointed in the merger," his son Robert was quoted later, "but he felt it was necessary to the survival of Hudson, and he had a keen regard for the interests of Hudson stockholders. He was an extremely faithful person." The accompanying newspaper report continued: "The last car

The 1957 Hornet Hollywood hardtop.

[Barit] owned was a Hudson built by AMC. When that car went out of service, A. E. Barit never bought another."

Some sources state that Barit left the board in anger over the decision to drop the Hudson name. Roy D. Chapin, Jr., doubts that, reiterating that Barit was always a realist: "We ran Hudson and Nash Metropolitans and Ramblers for a year—it was a charade. They were basically the exact same automobiles, and the decision really was one that said we've got to spend our money and our effort and our concentration on the Rambler because we haven't got enough dough to update the big Nashes and the big Hudsons." A rumor floated that Ramblers would diverge into very different Hudson and Nash types using the same basic shell, but money didn't exist for that either. So the final decision, as Chapin puts it, was, "Let's get it all behind one name and not try to advertise two different automobiles when they really are one. That's a simple explanation for a basic corporate decision that said we've got to put this whole thing together into one package."

Hornet dashboard for 1957.

For awhile the separate dealer organizations set up by terms of the original merger continued, but this, too, gradually evolved into all-Rambler, all-AMC agencies, which was more efficient. Cost and efficiency were also behind the 1958 abandonment of the Jefferson Avenue plant, which further obliterated the memory of Hudson when it was torn down that year and the ground used for a parking lot. "By fateful irony," Maurice Hendry wrote, ". . .all manner of marque jostle daily for space on the very site of what was once a respected competitor's headquarters. No doubt the majority of their owners never realize that cars were once made, not merely parked, right at that spot."

The abandonment of Jefferson "was just sheer economics," Roy Chapin says sadly. "I hated to see it happen, but the plant for twenty years preceding the merger—ten years anyway—was not maintained. All that was done to it was a little paint put on a couple of floors. It was really bad, by our standards today. Even on our older plants, we spend to keep them maintained. Everything works and they're not all beautiful but the quality's good. Hudson had a work space problem, very poor efficiency, a lot of people that had been there for years and had a lot of seniority. They had a very strong union. There were a lot of good reasons to get out of that plant."

Chapin is also positive that the idea of putting a Nash or Hudson nameplate on the Ambassadors was never seriously considered: "Presumably [Rambler] would appeal to the younger people. A lot of the same proposals got to us with the Javelin. After years of Classics and Ambassadors and Rambler Americans the decision was to try to take a shot at something different. More recently in that case it was a very useful and profitable thing."

Romney's decision to put his all into the Rambler, and the corollary decision to stick to that name instead of Nash and Hudson, was obviously the right one. Backed by aggressive advertising and a public mood that favored sensibly sized cars instead of behemoths, the Rambler went from strength to strength, production figures soaring. In 1957, his company turned the corner. Between 1957 and 1958, American Motors jumped from twelfth in the industry with 114,000 units to sixth with over 217,000. By 1960, when the Big Three turned in disbelief to find this spunky independent breathing down their corporate necks, Rambler had displaced Plymouth from its traditional number three spot and had built nearly half a million cars. In the wave of confidence and optimism that followed these notable achievements, it can surely be forgiven that AMC somehow managed to forget the loss of the marque that gave us the Super Six, the Essex and the Terraplane, the Commodore and the mighty Hornet.

Thus it came as a shock, and a poignant reminder, when a brand new 1970 American Motors car came out proudly bearing the name Hornet. Was it much like the Hornets of old? Well, yes and no. It wasn't the ponderous, over-engineered, beautifully crafted car the old Hornet had been, and it didn't pound over any kind of surface with equal ease, or run a 308-cubic-inch L-head six. It was compact, in the idiom of the time, and it usually came with a much tamer, overhead valve six. On the other hand, it handled well, used a unit body, and *could* be ordered with a potent V-8 that displaced just about the same cubic inches as the Hornet of 1951 through 1954. Relating it to this ancestor isn't easy, but it wasn't altogether unrelated either. Evidently the first Hornet had taught people something about the way a car ought to be built, and the way it should perform—if only enroute to the local A&P. And the new Hornet's 1970-75 sales were nearly 600,000 units.

There's justice in that.

Hudsons and Nashes for 1957, side by side at the Kenosha production line.

CHAPTER 7

Why Hudson Failed

THE DECLINE AND ultimate demise of Hudson, as a company and a make, is not a particularly hard phenomenon to explain. Did it have to happen? Very possibly. No matter what mistakes Hudson itself made, and there were several glaring ones, the temper of the times was against the firm at least from the dawn of the fifties—ironically so, considering that its most advanced product appeared in the 1951 Hornet.

Independents have existed for decades in the automobile business, and one still exists, because they provide products that the big manufacturers do not. They keep out of direct line-for-line and price competition and try to offer different styling or engineering, or features buyers can't find in the high-volume giants. Some of them, like American Motors today, were quite successful in following such a course. Packard, for example, was the dominant American luxury car from 1921 until the mid-thirties, because it built a car of unquestioned integrity and the very highest quality when such vehicles did not greatly interest the Big Three, who were fighting among themselves for the middle- and low-priced markets. Packard also brought the industry clever innovations almost from the day it was founded—the steering wheel, the H-patterned gear shift, spiral bevel gears, the first V-12 in regular production, the first successful straight eight and torsion-level ride.

It wasn't hard for the independents to survive before the war. Even Studebaker—in very difficult Ford-Chevy market territory—managed to do alright after it recovered from the depression, by introducing its inexpensive, fuel-sipping Champion in 1939. Nash was a pioneer in heating and ventilating, and offered 'liveability' through seats that converted into beds and coil spring independent front suspension. Hudson stressed high performance; its Terraplane was one of the fastest cars in its class in the world.

Even after the war, Kaiser-Frazer started from scratch and made progress by introducing the first straight-through fenderline, unique variations of the four-door sedan with hatchbacks and soft-tops, and an unprecedented variety of interior colors and materials. In 1948 Hudson fielded a product nobody else was even close to in terms of concept, design and roadability. Studebaker—not GM, Ford or Chrysler—was the first prewar manufacturer to come out with a brand new car after V-J Day, proving that a smaller company can often change direction faster than a large one.

In the last half of the forties the independents, like all manufacturers, could sell literally anything they built. Hudson, for example, could probably have kept right on selling the 1946 Commodore, without a single styling change, through 1950. Unfortunately, with the exception of George Mason, none of the independents had the vision to see that the seller's market wouldn't last forever, and (excepting Nash, which restyled in 1949 *and* 1952 and developed the Rambler and the Nash-Healey) failed to act accordingly. That was their first mistake. It applied to them all, even Kaiser-Frazer: Henry Kaiser and Joe Frazer actually split over Henry's determination to build twice as many cars for 1949 as he could sell.

What the independents didn't seem to realize was the enormous size differential between themselves and the Big Three. Said Mike Lamm in *Special-Interest Autos* in 1973: "It costs as much for a small company to retool a new car as it does a big one. Chevy or Ford, though, can amortize that tooling cost over, say, a million cars, while [an independent] is lucky to produce a tenth of that number." Mike also pointed out that it cost an independent as much to buy a one-page ad in *Life* as it did for one of the Big Three—there were no discounts for low-production. If you build a million cars that ad may cost you a cent or two a car, but if you build only 100,000 the cost is ten cents.

Another problem was the quantity of plants. GM, Ford and Chrysler had assembly plants in various locations throughout the country. They could ship knocked-down cars or raw materials to them and turn out cars for the local area, holding their transportation costs to a minimum. Companies like Hudson had hardly any plants outside the midwest, and had to foot the bill for delivering their cars all over the country.

Undoubtedly the independents suffered drastically from the 'Ford Blitz' of 1953. The overproduction of Ford and GM, in their strident efforts to bury each other, buried only the independents. Studebaker in particular suffered from the Ford-GM onslaught (in 1953 Studebaker built 186,000 cars; in 1954 85,000). This was probably unintentional. But it happened.

Then there was the federal government whose actions were, as usual, well-intended but in the end devoid of much benefit. In its postwar price fixing and loan controls, the Federal Reserve Board and other agencies crippled the ability of car manufacturers to do business, just around the time (1949-50) that the Big Three were getting back into stiff competition with all-new, fully postwar products. The FRB's eighteen month limit on credit purchases affected all dealerships equally, but percentage-wise it was a much bigger problem to companies like Hudson than companies like Ford. Then the Defense Department extended a helping hand during the Korean conflict, providing generous contracts to firms such as Hudson for military equipment—only to yank the carpet out from under them without any warning as the political situation stabilized. One can argue that management should have seen the end of the Korean War coming, but as late as 1953 most people had no real idea when it would end, and the government was crying louder than ever for deliveries of aircraft components and other military hardware. Packard set up a jet-engine complex at enormous expense, only to have the government cancel its contract without any warning in 1954.

Looking at the independents in general after the war, it is possible to visualize two ways out of the coming crisis. First, had George Mason been able to persuade others to adopt his line of thinking, a huge amalgam of Hudson, Nash, Packard and Studebaker might have been able to compete head-to-head with the Big Three on an equal basis. We would have then had four major companies in the automobile industry instead of three majors and four minors. (Kaiser-Frazer, and Kaiser-Willys after them, were considered maverick outfits run by West Coasters who didn't know much about automobiles.) The second choice was to continue to take daring chances on a product in order to outflank the big companies by producing something totally different, yet equally salable.

Inevitably one is again forced to admire the foresight of George Mason: Nash-Kelvinator adopted *both* of these programs! The grand scheme of a giant merger was long-term, in the meantime Nash would continue to make progress on its own with the Rambler, and the big Nashes with their many unique features. Said *Fortune* in 1954, "Nash's Rambler (in contrast to the Henry J), was fully equipped and shrewdly offered at a price that just topped the Big Three's low-priced makes. It has found a place as a second or utility car among prosperous suburbanites. The Rambler, now established, is being offered in 1955 models without full equipment at prices under Chevrolet and Ford. The Rambler . . .never gets into direct price competition with them." Given these actions on the part of Mr. Mason, is it any wonder that Nash alone, of all the independents, managed to create a product on which it could survive into the sixties?

As good as the Hornet and other Step-down models were, it must be admitted that Hudson management was nowhere near as far-sighted. In the 1948 Hudson they created a wonderful automobile—solid, durable, roadable. In the 1951 Hudson Hornet they added the finest L-head six that had ever been produced, with performance that was startling. But the Step-down was almost impossible to alter, and annual face-lifts or restyles were the game a company *had* to play in order to remain in competition during the fifties. To repeat Roy Chapin's comment, "The only place you could alter it was above the beltline, without going into extensive retooling."

During this author's Chapin interview, AMC assistant public relations director John Conde asked if Hudson's inability to provide a station wagon on the Step-down chassis, or air conditioning, had any bearing on its sales problems. "Yes," said Mr. Chapin, "as symptomatic of the old question of the chicken and the egg. If you don't have enough money to do something, and do it right, and if you haven't learned to specialize in a given thing and concentrate your time and effort, sooner or later you find you just can't do everything. The station wagon was such a major structural change that they couldn't afford to do it. The air conditioning was like a lot of things. Many times things that Hudson did were because they got to the point where they just had to do them. There was a lot of the attitude, 'Jeez, we gotta do this now,' instead of saying, 'well, we'd better do this because two years from now we're going to need the damn thing.' If any basic criticism exists, that was it. They were usually reacting, rather than anticipating. It was not all voluntary—they'd liked to have done the other, but they were not equipped, in some cases emotionally or in other cases financially, to take the risk."

Hudson's seeming inability to produce, and a statement in one article that its actual breakeven point eventually rose higher than its maximum plant output, caused the author to query Mr. Chapin on these subjects. "I remember when we got below forty cars an hour we were in trouble," he said. "That's 320 a day times 230 days, 73,600. I think 75,000 would be a close figure to Hudson breakeven. I don't know where [the theory of breakeven being higher than maximum output] comes from. They could do 175,000 cars a year on one shift, two shifting—there was nothing to it. If they could have sold them they'd have made a lot of money."

Another factor that should be considered in analyzing Hudson's decline is its typically weak dealer situation—typically for an independent in those days. Stuart Baits feels this is a significant reason behind the company's failure: "In the middle fifties, all the remaining independent dealers were going over to the Big Three, who could make more attractive offers. This is a perfectly natural development in business and it had eaten up all the other [independent] makers down to the last four already. When we had an unusually successful dealer he could not help but wonder about switching, even if he had not been approached." And Hudson had some unusually *unsuccessful* dealers too. Like all independents, except possibly Nash (again!), its volume-per-dealer was low and the high volume agents naturally tended to move toward Big Three franchises.

We have already mentioned herein the effect of closing down Hudson distributorships. There are mixed feelings on this. Some observers say this step strengthened both Hudson and its outlets by eliminating the profits of a middleman, while others (James J. Nance is one) feel that a good distributor "took care of his dealers" and insured their loyalty to the factory. In some ways, however, Hudson had to eliminate its distributorships, because there were growing instances of a distributor deserting to another manufacturer, and taking a ring of client dealers with him. This increased considerably in the early fifties when sales departments of large companies (especially Ford) went on raiding parties, zeroing in on Chrysler and independent agencies in particular. Doubtless, all part of that reckless Ford attempt to head-off General Motors. The independents had a lot to dislike about their competitors in Dearborn . . .

Looking back—and it is always easy to look back—one can see finally that the Hudson Jet was the single greatest mistake the company ever made. The reasons why have been exhaustively outlined in Chapter 4 and will not be reiterated here. Productwise it was almost exactly wrong, and its timing was guaranteed to make it a failure. This miscalculation is magnified by the fact that the $16 million it took to tool up for the Jet could have more than paid for that V-8 engine Hudson needed so badly in its big cars.

The importance of a V-8 in any automotive lineup around 1953 or 1954 cannot be underestimated. This does not imply that the Hornet six was any less an engine—it was a great engine, something to be proud of. But great engines, even great cars, do not necessarily guarantee success—which is in the end tied to that unfathomable marketplace composed of us fickle, usually unpredictable, often illogical human beings. We are all that, we cannot deny it. Whether a Hornet would keep up with a V-8 in a straight line, and lose it over a twisty bit of road, was not the point. What mattered was that all people saw when they cracked open the glistening hood were six spark plugs. That, and a body that could easily be identified with as one Hudson offered in 1948, made the Hornet unviable in the mid-fifties marketplace.

And so it comes down to management, as it always must, for management in the end is the major element in the success and failure of a corporation. In management's ability to predict and satisfy the market lies survival, or bust. In that, Hudson management failed. That the company nevertheless built automobiles that are now remembered—and collected—to a far greater degree than the more 'successful' ones, is a little added bonus, for which enthusiasts may be thankful.

Appendix I

Hudson Statistical Summary, 1945-1954

YEAR	NET SALES	NET EARNINGS
1945	71,740,601	673,248
1946	120,715,415	2,382,641
1947	159,514,329	5,763,352
1948	274,728,638	13,225,923
1949	259,597,308	10,111,219
1950	267,219,750	12,002,274
1951	186,050,833	(1,125,210)
1952	214,873,982	8,037,848
1953	192,846,084	(10,411,060)
1954 (to 4-30-54)	28,685,014	(6,231,570)

(figures in parenthesis indicate loss)

Appendix II

Hudson Production Figures, 1946-1957

PRODUCTION 1946	
'51' SUPER, SIX-CYLINDER	61,787
'52' COMMODORE, SIX-CYLINDER	17,685
'53' SUPER, EIGHT-CYLINDER	3,961
'54' COMMODORE, EIGHT-CYLINDER	8,193
'58' COMMERCIAL SUPER, SIX-CYLINDER	3,374
TOTAL	95,000

PRODUCTION 1947	
'71' SUPER, SIX-CYLINDER	49,276
'72' COMMODORE, SIX-CYLINDER	25,138
'73' SUPER, EIGHT-CYLINDER	5,076
'74' COMMODORE, EIGHT-CYLINDER	12,593
'78' SUPER, SIX-CYLINDER	2,917
TOTAL	95,000

PRODUCTION 1948	
'81' SUPER, SIX-CYLINDER	49,388
'82' COMMODORE, SIX-CYLINDER	27,159
'83' SUPER, EIGHT-CYLINDER	5,338
'84' COMMODORE, EIGHT-CYLINDER	35,315
TOTAL	117,200

PRODUCTION 1949	
'91' SUPER, SIX-CYLINDER	91,333
'92' COMMODORE, SIX-CYLINDER	32,715
'93' SUPER EIGHT-CYLINDER	6,365
'94' COMMODORE, EIGHT-CYLINDER	28,687
TOTAL	159,100

PRODUCTION 1950

Model	Production
'500' PACEMAKER, SIX-CYLINDER	39,455
'50A' PACEMAKER, SIX-CYLINDER	22,297
'501' SUPER, SIX-CYLINDER	17,246
'502' COMMODORE, SIX-CYLINDER	24,605
'503' SUPER, EIGHT-CYLINDER	1,074
'504' COMMODORE, EIGHT-CYLINDER	16,731
TOTAL	121,408

PRODUCTION 1951

Model	Production
'4A' PACEMAKER, SIX-CYLINDER	34,495
'5A' SUPER, SIX-CYLINDER	22,532
'6A' COMMODORE, SIX-CYLINDER	16,979
'7A' HORNET, SIX-CYLINDER	43,666
'8A' COMMODORE, EIGHT-CYLINDER	14,243
TOTAL	131,915

PRODUCTION 1952

Model	Production
'4B' PACEMAKER, SIX-CYLINDER	7,486
'5B' WASP, SIX-CYLINDER	21,876
'6B' COMMODORE, SIX-CYLINDER	1,592
'7B' HORNET, SIX-CYLINDER	35,921
'8B' COMMODORE, EIGHT-CYLINDER	3,125
TOTAL	70,000

PRODUCTION 1953

Model	Production
'1-2C' JETS, SIX-CYLINDER	21,143
'4-5' WASP, SIX-CYLINDER	17,792
'7C' HORNET, SIX-CYLINDER	27,208
TOTAL	66,143

PRODUCTION 1954

Model	Production
'1-2-3D' JETS, SIX-CYLINDER	14,224
'4-5D' WASP, SIX-CYLINDER	11,603
'6-7D' HORNET, SIX-CYLINDER	24,833
TOTAL	50,660

PRODUCTION 1955

Model	Production
HUDSON RAMBLER, 100″ WB	
Deluxe Delivery Van	21
Deluxe Business Coupe	34
Super Suburban	1,335
Super Two-Door sedan	2,970
Custom Country Club	1,601
TOTAL	5,981
HUDSON RAMBLER, 108″ WB	
Four-Door sedan	7,210
Cross Country	12,023
TOTAL	19,233
WASP	
Four-Door sedan	5,551
Hollywood hardtop	1,640
TOTAL	7,191
HORNET	
Four-Door sedan (6)	5,357
Four-Door sedan (V8)	4,449
Hollywood hardtop (6)	1,554
Hollywood hardtop (V8)	1,770
TOTAL	13,130
TOTAL ALL MODELS	45,535

PRODUCTION 1956

Model	Production
WASP	
Four-Door sedan	2,519
HORNET SPECIAL	
Four-Door sedan	1,528
Hollywood hardtop	229
TOTAL	1,757
HORNET	
Four-Door sedan (6)	3,022
Four-Door sedan (V8)	1,962
Hollywood hardtop (6)	358
Hollywood hardtop (V8)	1,053
TOTAL	6,395
TOTAL ALL MODELS	10,671

PRODUCTION 1957

Model	Production
Custom Four-Door sedan	2,256
Super Four-Door sedan	1,103
Custom Hollywood hardtop	483
Super Hollywood hardtop	266
TOTAL CUSTOM	2,739
TOTAL SUPER	1,369
TOTAL ALL MODELS	4,108**

**The Hudson-Essex-Terraplane Club's *White Triangle News* for March/April 1993 provides the 1957 figures above, along with many other interesting details. For instance, 242 of the '57 Hudsons were righthand drive models; 1,064 Customs had three-tone paint and 1,193 two-tone; 1,222 Supers were two-toned; another 256 were shipped "in the white" or painted special-order colors, leaving only 373 in solid colors. The most popular solid was black, while the favorite three-tone combination was Cinnamon Bronze, Sierra Peach and Frost White (on 387 Customs). The most popular optional equipment was power brakes (3,876 cars), power steering (3,321), radio (3,121), clock (2,830), tinted glass (1,901) and windshield washer (1,719); 415 1957 Hudsons had air conditioning.

Appendix III

Hudson Hornet Racing Record, 1951-1955

1951
Daytona Beach, FL 2-11 160 miles
Marshall Teague, first

Gardena, CA 4-8 100 miles
Marshall Teague, first

Phoenix, AZ 4-22 150 miles
Marshall Teague, first

Canfield, OH 5-30 100 miles
Marshall Teague, first

Gardena, CA 6-30 100 miles
Lou Figaro, first

Grand Rapids, MI 7-1 100 miles
Marshall Teague, first
Dick Rathmann, second

Des Moines, IA 9-4 125 miles
Dick Rathmann, first

Darlington, SC 9-4 500 miles
Herb Thomas, first
Jesse James Taylor, second

Langhorne, PA 9-16 150 miles
Herb Thomas, first

Charlotte, NC 9-23 150 miles
Herb Thomas, first

Occoneechee, NC 10-6 150 miles
Herb Thomas, first
Lennie Tippet, second

Jacksonville, FL 11-4 100 miles
Herb Thomas, first
Jack Smith, second

Atlanta, GA 11-11 100 miles
Tim Flock, first

1952
West Palm Beach, FL 1-20 100 miles
Tim Flock, first

Dayton Beach FL 2-10 150 miles
Marshall Teague, first
Herb Thomas, second

Jacksonville, FL 3-16 100 miles
Marshall Teague, first
Herb Thomas, second
Tim Flock, fourth
Tommy Moon, fifth

Gardena, CA 3-23 100 miles
Danny Letner, first
Lou Figaro, second

Tampa, FL 3-30 100 miles
Jim Thompson, first

North Wilkesboro, NC 3-30 125 miles
Herb Thomas, first

Martinsville, WV 4-6 100 miles
Dick Rathmann, first
Perk Brown, third

Columbia, SC 4-12 100 miles
Buck Baker, first
Dick Rathmann, third
Joe Eubanks, fourth

Macon, GA 4-27 100 miles
Herb Thomas, first

Langhorne, PA 5-4 150 miles
Dick Rathmann, first
Tim Flock, second

Darlington, SC 5-10 100 miles
Dick Rathmann, first
Tim Flock, second

Toledo, OH 5-11 100 miles
Marshall Teague first

Dayton, OH 5-18 100 miles
Dick Rathmann, first

Gardena, CA 5-25 100 miles
Lou Figaro, first

Canfield, OH 5-30 100 miles
Herb Thomas, first
Bob Moore, fourth
Joe Eubanks, fifth

San Diego, CA 5-30 500 laps
(¼ mile track)
Danny Letner, first

Toledo, OH 6-1 100 miles
Tim Flock, first
Dick Rathmann, second

Augusta, GA 6-1
Tommy Moon, second

Occoneechee, NC 6-8 100 miles
Tim Flock, first
Dick Rathmann, third

Charlotte, NC 6-15 100 miles
Herb Thomas, first
Tim Flock, second
Dick Rathmann, fifth

Detroit, MI 6-29 250 miles
Tim Flock, first
Buddy Shuman, second
Herb Thomas, third

Niagara Falls, ONT 7-1 100 miles
Buddy Shuman, first
Herb Thomas, second

Toledo, OH 7-4 50 miles
Marshall Teague, first
Red Hamilton, second
Rodney Clark, third
Walter Minx, fourth

Owego, NY 7-4 100 Miles
Tim Flock, first
Herb Thomas, second
Dick Rathmann, third
Buck Sager, fourth

Williams Grove, PA 7-6 100 miles
Marshall Teague, first
Jack McGrath, second
Frank Luptow, third

Monroe, MI 7-6 100 miles
Tim Flock, first
Herb Thomas, second

Milwaukee, WI 7-13 150 miles
Marshall Teague, first
Buck Barr, fourth
Red Hamilton, sixth
Rodney Clark, eighth

Gardena, CA 7-12 100 laps (50 miles)
Lou Figaro, first

South Bend, IN 7-20 100 miles
Tim Flock, first
Herb Thomas, third

Dayton, OH 7-27 100 miles
Marshall Teague, first
Red Hamilton, fourth

Richmond, VA 8-3 100 miles
Jack McGrath, first
Marshall Teague, second

Williams Grove, PA 8-10 50 miles
Jack McGrath, first
Red Hamilton, second

Rochester, NY 8-15 100 miles
Tim Flock, first
Herb Thomas, second
Dick Rathmann, third

Asheville, NC 8-17 100 miles
Bob Flock, first
Tim Flock, second
Herb Thomas, third
Gene Comstock, fourth

Milwaukee, WI 8-22 100 miles
Frank Luptow, first
Red Hamilton, second
Jack McGrath, third
Marshall Teague, fourth

Terre Haute, IN 8-24 100 miles
Frank Luptow, first
Buck Barr, second
Marshall Teague, fifth

Milwaukee, WI 9-7 200 miles
Marshall Teague, first
Frank Luptow, second

Atlanta, GA 9-21 100 miles
Marshall Teague, first
Frank Mundy, second
Red Byron, third

Dayton, OH 9-21 100 miles
Dick Rathmann, first
Herb Thomas, fifth

Springfield, IL 9-28 100 miles
Frank Munday, first

Wilson, NC 9-28 100 miles
Herb Thomas, first
Dick Rathmann, fifth

Bakersfield, CA 10-11 25 miles (PRC)
Danny Letner, first

Pomona, CA 10-12 100 miles
Jack McGrath, first
Andy Linden, third
Bob Christy, fourth
Marshall Teague, fifth

Martinsville, VA 10-19 100 miles
Herb Thomas, first
Tim Flock, fourth
Johnny Patterson, fifth

Klamath Falls, OR 10-19 100 miles
George Amick, first

North Wilkesboro, NC 10-26 100 miles
Herb Thomas, first
Don Thomas, third
Tim Flock, fourth
Dick Rathmann, fifth

Atlanta, GA 11-16 100 miles
Herb Thomas, first
Perk Brown, third
Tim Flock, fourth
Joe Eubanks, sixth
Jack Smith, seventh

West Palm Beach, FL 11-30 100 miles
Herb Thomas, first
Perk Brown, third

Gardena, CA 12-7 100 miles
Frank Mundy, first
Jack McGrath, second

1953

Gardena, CA 2-1 100 miles
Dick Meyers, first

Gardena, CA 2-22 100 miles
Frank Mundy, first
Fred Steinbronner, third

Spring Lake, NC 3-8 100 miles
Herb Thomas, first
Dick Rathmann, second

Gardena, CA 3-22 100 miles
Dick Meyers, first

Phoenix, AZ 3-29 100 miles
Jack McGrath, first
Ebe Yoder, second
Jimmy Bryan, third
Frank Mundy, fifth

North Wilkesboro, NC 3-29 125 miles
Herb Thomas, first
Dick Rathmann, second

San Mateo, CA 4-25 250 laps
Lou Figaro, first

Gardena, CA 4-26 100 miles
Frank Mundy, first
Jack McGrath, second
Fred Steinbronner, third

Macon, GA 4-26 100 miles
Dick Rathmann, first
Herb Thomas, second
Tim Flock, sixth

Culver City, CA 5-9 100 laps
Dick Meyer, first
Bill Galdrisi, second (1950 Hudson)

Toledo, OH 5-10 100 miles
Frank Mundy, first
Marshall Teague, second
Johnny Mantz, third
Bill Taylor, fourth

Hickory, NC 5-16 100 miles
Tim Flock, first
Joe Eubanks, second
Dick Rathmann, fifth

Columbus, OH 5-24 100 miles
Herb Thomas, first
Dick Rathmann, second
Pop McGinnis, fifth

Raleigh, NC 5-30 300 miles
Fonty Flock, first
Tim Flock, second
Herb Thomas, third
Dick Rathmann, fifth

Gardena, CA 5-30 250 miles
Lou Figaro, first

Youngstown, OH 5-30 200 laps
Jim Romine, first

Pensacola, FL 6-14 100 miles
Herb Thomas, first
Dick Rathmann, second
Tim Flock, fifth
Joe Eubanks, sixth

Williams Grove, PA 6-14 50 miles
Jack McGrath, first
Marshall Teague, second
Frank Mundy, third
Bill Taylor, fourth

Wellington, OH 6-14 200 laps
Jim Romine, first
Bucky Sager, second

Langhorne, PA 6-21 200 miles
Dick Rathmann, first
Herb Thomas, fourth

High Point, NC 6-26 100 miles
Herb Thomas, first
Dick Rathmann, second
Joe Eubanks, third

Gardena, CA 6-27 200 laps
Lou Figaro, first
Danny Letner, fourth (1951 Hudson)

Wilson, NC 6-28 100 miles
Fonty Flock, first
Dick Rathmann, second
Herb Thomas, third
Joe Eubanks, fourth

Rochester, NY 7-3 100 miles
Herb Thomas, first
Dick Rathmann, second
Tim Flock, fourth

Heidelberg, PA 7-4 50 miles
Frank Mundy, first
George Fleming, seventh
Fred Wheat, ninth

Morristown, NJ 7-10 100 miles
Dick Rathmann, first
Herb Thomas, second

Atlanta, GA 7-12 100 miles
Herb Thomas, first
Dick Rathmann, second
Joe Eubanks, fourth

Milwaukee, WI 7-12 150 miles
Frank Mundy, first
Jack McGrath, fourth
Johnny Mantz, sixth

Rapid City, SD 7-22 100 miles
Herb Thomas, first
Dick Rathmann, second
Fonty Flock, third

Illiana Speedway 7-25 100 miles
Jack McGrath, first

North Platte, NE 7-26 100 miles
Dick Rathmann, first
Herb Thomas, second

Dayton, OH 7-26 200 laps
Jim Romine, first
Bucky Sager, second
Russ Hepler, third

Davenport, IA 8-2 100 miles
Herb Thomas, first
Dick Rathmann, fourth
Fonty Flock, fifth
Bill Harrison, sixth

Winchester, IN 8-9 100 miles
Marshall Teague, first
Frank Mundy, second

Salem, IN 8-16 25 miles
Marshall Teague, first
Sam Hanks, second
Frank Mundy, third

Asheville, NC 8-16 100 miles
Fonty Flock, first
Herb Thomas, second

Milwaukee, WI 8-23 100 miles
Sam Hanks, first
Marshall Teague, second
Frank Mundy, third
John Mantz, sixth (1952 Hornet)

Norfolk, VA 8-23 100 miles
Herb Thomas, first
Fonty Flock, second
Dick Rathmann, fourth

Hickory, NC 8-29 100 miles
Fonty Flock, first
Herb Thomas, second
Joe Eubanks, third

Syracuse, NY 9-10 100 miles
Sam Hanks, first
Frank Mundy, second
L. Baltes, fourth

Fort Wayne, IN 9-13 100 laps
Marshall Teague, first
Sam Hanks, second
Frank Mundy, third

Dayton, OH 9-13 250 miles
Iggy Katona, first
Jim Romine, second

Langhorne, PA 9-20 250 miles
Dick Rathmann, first
Herb Thomas, second
Jim Reed, fourth

Bloomsburg, PA 10-3 100 miles
Herb Thomas, first
Dick Rathmann, second

Wilson, NC 10-4 100 miles
Herb Thomas, first
Fonty Flock, third
Dick Rathmann, fifth

Gardena, CA 10-11 100 miles
Danny Letner, first
Lou Figaro, second

1954

Gardena, CA 1-31 100 miles
- George Fleming, first
- Lou Figaro, third
- Marshall Teague, fifth
- George Amick, seventh
- Fred Steinbroner, tenth

West Palm Beach, FL 2-7 100 miles
- Herb Thomas, first

Jacksonville, FL 3-7 100 miles
- Herb Thomas, first
- Fonty Flock, second
- Joe Eubanks, fourth

Gardena, CA 3-14 50 miles
- Troy Ruttman, first
- Marshall Teague, second
- Frank Mundy, fifth
- Fred Steinbroner, sixth

Atlanta, GA 3-21 100 miles
- Herb Thomas, first
- Dick Rathmann, third
- Fonty Flock, fourth

Oakland, CA 3-28 125 miles
- Dick Rathmann, first

Savannah, GA 3-28 100 miles
- Al Keller, first
- Don Thomas, fourth
- Joe Eubanks, fifth
- Herb Thomas, tenth

Phoenix, AZ 3-28 100 miles
- Marshall Teague, first
- Frank Mundy, third
- Bill Parks, fifth

North Wilkesboro, NC 4-4 100 miles
- Dick Rathmann, first
- Herb Thomas, second
- Joe Eubanks, third

Hillsboro, NC 4-18 100 miles
- Herb Thomas, first
- Don Thomas, second
- Dick Rathmann, fourth

Langhorne, PA 5-2 150 miles
- Herb Thomas, first
- Al Keller, second
- Dick Rathmann, third
- Joe Eubanks, fourth
- Paul Petitt, fifth

Knoxville, TN 5-2 50 miles
- Frank Mundy, first
- Marshall Teague, second
- Bob Peterson, thrird
- Irwin Kirble, fifth
- Lou Figaro, sixth
- George Fleming, seventh

Raleigh, NC 5-29 250 miles
- Herb Thomas, first
- Dick Rathmann, second
- Al Keller, sixth
- Bill Blair, tenth

Hickory, NC 6-19 100 miles
- Herb Thomas, first
- Dick Rathmann, fourth
- Junior Johnson, fifth
- Joe Eubanks, eighth

Williams Grove, PA 6-27 100 miles
- Herb Thomas, first
- Dick Rathmann, second
- Joe Eubanks, fourth
- Ralph Ligouri, eighth

Spartanburg, SC 7-3 100 miles
- Herb Thomas, first
- Joe Eubanks, fifth (1951 Hudson)
- Dick Rathmann, seventh
- Arden Mounts, eighth

Asheville, NC 7-4 100 miles
- Herb Thomas, first
- Dick Rathmann, third
- Joe Eubanks, sixth (1952 Hornet)

Willow Springs, IL 7-10 100 miles
- Dick Rathmann, first
- Herb Thomas, second

In September 1955, *Motor Trend* reported the following NASCAR Grand National Standings:

Pos.	Make	Cars	Points	Percentage
1.	Oldsmobile	143	414	29.0
2.	Chrysler	57	235	41.2
3.	Hudson	78	130	16.7
4.	Chevrolet	50	85	17.0
5.	Dodge	39	68	17.4
6.	Mercury	21	46	21.9

Although none of the racing Hudsons were later than 1954 models, the cars continued to win against such powerful rivals as the new Chevrolet and Dodge V-8's.

Appendix IV

Compiled by Jeffrey I. Godshall and Autoenthusiasts International

Hudson Literature, 1946-1957

1946:
Catalog, 9×7, full line w/convertible, color, 20p. 5-46
Catalog, 14×11, full line, color, 20p., 11-45
Catalog 9×7, full line, color, 20p., 11-45
Folder, 8½×11, full line, sepia&w, 9-45
Folder, 9×6¼, pick-up, b&w&yellow
Catalog, 5½×8½, "Jubilee Parade of Progress", b&w&red, 12p., 29 5-46

1947:
Catalog, 9×7, full line, color, 20p., reprinted 3-47

1948:
Catalog, 8½×11, full line, color covers, w/brown&yellow pgs., 24p. w/foldout covers, 10-47
Catalog, 5×7, "Stepping Down", b&w&red, 16p. w/foldout covers, 6-48
Catalog, 14×11, full line, color, 24p. w/foldout covers, 11-47
Folder, 11×8, convertible, color, 8-48
Folder, 8½×11, full line, color, 10-47
Catalog, 8½×11, full line, color covers, w/brown&yellow pgs., 24p. w/foldout covers, 11-47
Mailer, 8×5, sedan, b&w&red, 3-48

1949:
Folder, 6×9, Mechanix Illustrated test, b&w&blue
Catalog, 14×11, full line, color, 24p. + foldout, AM 497 1-49
Catalog, as above, 5-49
Folder, 8½×11, full line, color, AM 499 1-49
Folder, as above, 5-49
Catalog, 8½×11(saver), full line, color, 24p., 5-49
Catalog, 8½×11, "Features", full line, sepia&yellow&w w/color covers, 28p., AM 498 5-49
Catalog, 8½×11, "Features", color, AM 498 12-48
Catalog, as above, 1-49

1950:
Catalog, 10×8, full line, color, 28p., AM 5011 4-50
Mailer, 8½×5½, full line, color, AM 5017 1-50
Catalog, 10×8, full line, color, 28p., AM 5011 1-50
Catalog, 9×11, full line, b&w, 24p., 5013 1-50
Folder, 8½×11, Pacemaker, color, AM 5002 11-49

1951:
Catalog, 10×8, full line, color, 32p., AM 5103 9-50
Catalog, 10×8, full line, color, 32p., AM 5103-r 3-51
Catalog, 10×8, Hornet, color, 8p., AM 5104 9-50
Folder, 11×16, Racing Hornet, b&w, AM 5147 6-51
Booklet, 5×7, StepDownDesign, b&w&red, 16p. plus covers, AM 5130 4-51
Booklet, 6×9, Hornet Road test reprint, b&w, AM 5211 12-51

1952:
Catalog, 3½×6½, Hornet racing, b&w, 11p., AM 5230 3-52
Catalog, 12×8½, full line, color, 28p., AM 5203 12-51
Catalog, 11×15½, full line, color, 8p., AM 5204 12-51

1953:
Card, 5½×3, Jet, color
Folder, 11×9½, Hudson, color, AM 5302 11-52 (incorrect Hornet illust. on cover)
Folder, 11×9½, Hudson, color, AM 5302 11-52 (correct Hornet illust.)
Folder, 11×9½, Hudson, color, AM 5302 3-53
Catalog, 10½×15, Jet, color, 8p., AM 5301 11-52
Folder, 8½×11, Hudson Dealer News showing "Italia", b&w w/color covers, Sept. 53

1954:
Catalog, 14×11, full line, color, 16p., AM 5401 9-53
Folder, 14×11, Hornet Special/Jet Club Sedan, b&w, 5-54 [AM 5438]
Card, 8×5, Hornet Special/Club Sedan, color AM 5447 5-54
Card, 8×5, Jet Family Club Sedan, color AM 5446 5-54
Catalog, 10½×15, full line, color, 8p., 9-53 AM 5402
Folder, 10×7½, access., sepia&w, SO-20 10-53
Catalog, 6×9, Power/engine options, b&w&red, AM 5414 12-53
Folder, 11×8½, Metropolitan, color, AM 5458 8-54
Rambler Folder-Mailer, "You Are Invited", 8×5, HM56-1572 10-55
Rambler Folder, 8×5½, Car Life road test, half-color
Rambler Magazine, 11×14, Nash News, No. 3'56, b&w&red, w/Rambler coming off assembly line w/mid-year trim
Rambler Catalog, 13×9, full line, color, 12p., R-56 5003
Rambler Folder, 11×6½, wagons, color, NSP-56 5199
Rambler Booklet, 8½×11, X-Ray, b&w, 36p., R-56 5149 3-56

1955.

Folder, 10×7½, access., part color
Folder, 3×6, air cond., b&w&red&blue, 1-55 HM-55-1069
Card, 8×5, Hollywood Hornet, color, 12-54 HM-55-1057
Card, as above, Hornet Sedan, 1-55
Card, as above, Wasp Hollywood, 1-55
Card, as above, Wasp Sedan, 1-55
Catalog, 14×11, Hornet/Wasp, color, 12p., 1-55 HM-55-1013
Folder, 10½×15, Hornet/Wasp, color, 1-55 HM-55-1021
Folder, as above 4-55
Folder, 8½×6, Announcement, color, w/envelope
Folder, 11×8½, Metropolitan, color NSP-55-1858 1 5-55
Folder, 7×6, Hudson Rambler, color
Rambler Catalog, 12×8, full line, color, 16p.
Rambler Folder, 9×10½, full line, color, NSP-55-1798 11-54
Rambler Folder, as above, 4-55
Rambler Catalog, 8×11, Comparison, b&w, 20p., NSP-55-2086 6-55
Rambler Catalog, 12×16, Comparison, b&w&red, 20p., NSP-55-2073 6-55
Rambler Booklet, 8½×11, Comparison, b&w, 8p. NT-55-1984
Rambler Folder, 8½×11, Cross Country, b&w&blue
Rambler Folder, 11½×7½, 'Why Every Woman . . .', b&w, 1-55 HM-55-1099

Note: Both Rambler and Metropolitian sold under Nash and Hudson nameplates.

1956:

Folder, $¼ Million Contest, b&w&red
Folder, 10×5½, access., (inc. Ramb.), brown&w, HAP-56-1575
Folder, 12×4½ "You Are Invited to See", color
Catalog, 15×10½, Hornet/Wasp/Ramb., color, 8p., 10-55 HM-56-1469
Catalog, as above w/Hornet Special, 4-56 HM-56-1763
Booklet, 8×11, X-Ray, b&w, 28p., 6-56 HM-56-1717
Folder-mailer, 5½×3, "Stand Out From Crowd', color, 10-55 HM-56-1464-A
Folder-mailer, as above, 'Drive Standout Car', color, 10-55 HM-56-1464-B
Metropolitan Folder, 12×9, '1500', b&w&green, M56-5274
Metropolitan Folder, as above, rev. (no mark)
Metropolitan Card, 8×5½, '1500' convert., b&w&green
Metropolitan Booklet, 4×6, 'Burning Money', b&w, 16p.

1957:

Folder, 10×10, Hornet, color, 9-56 HM-57-5780
Booklet, 8×11, X-Ray, b&w&red&blue, 24p.
Postcards, 5½×3, Hornet Custom hardtop/Super sedan, HM-57-5777-A&B

Index

Airfoam seats, 9
Allison, Bill, 48
American Motors Corporation
  Facilities
    Detroit, MI, 112, 122
    Kenosha, WI, 112, 123
    Milwaukee, WI, 112
  Formation of, 62, 96, 97, 106-112
American Society of Engineers' Merit Award, 64
Amick, George, 130, 131
Anderson, Edmund E., 116
Andrews, Robert F., 12, 15, 18, 20, 35-39, 40, 48
Andriesse, Albert H., 26
Armand, Harry, 50
Army and Navy 'E' Award, 13
Ascari, Alberto, 70
Atitt, R. A., 84
Automobile Manufacturers Association, 108

Badertscher, Dana, 34, 59
Baits, Stuart G., 11, 12, 26, 33, 34, 35, 37, 44, 46, 51, 58, 59, 66, 73, 100, 102, 103, 126
Baker, Buck, 129
Bald, Milton, 42
Baltes, L., 130
Barit, A. E., 11, 12, 13, 16, 17, 22, 23, 26, 29, 30, 33, 40, 43, 47, 48-50, 51, 64, 66, 77, 79, 80, 81, 82, 83, 87, 96, 99, 100, 105, 107, 108, 111, 112, 120-121
Barit, Robert, 100, 120-121
Barr, Buck, 129
Beguhn, E. J., 25
Behn, G. G., 26
Bentley, John, 86
Best, Al, 50
Bezner, F. O., 6, 11
Biddle & Smart Company, 9
Billingsley, Oliver, 76
Blair, Bill, 131
Bond, John, 76, 83, 85, 100
Bonneville, 9
Booz, Peter, 72
Boyce, Terry, 81
Boyd, Virgil, 96
Brady, J. J., 6
Brakes, descriptions of,
  Hydraulic, 9
  Rotary equalized, 9
Breech, Ernie, 88
Brough Superior, 10
Brown, Perk, 129, 130
Bryan, Jimmy, 130
Buchanan, Hershel, 68
Buege, A. F., 104
Bullard, Harold, 118
Byron, Red, 129

Caleal, Dick, 35
Calhoun, Howard, 25
Calvin, Bob, 69
Carrozzeria Touring, 98, 100, 101, 102, 103-104
Catlin, Russ, 58
Chalmers, Hugh, 6
Chapin, Roy D., Jr., 11, 12, 15, 26, 33, 48, 58, 69, 70, 78, 82, 87, 88, 98, 99, 100, 103, 104, 108, 110, 111, 121, 122, 125-126
Chapin, Roy D., Sr., 6, 9, 10, 11, 44
Chapin, Mrs. Roy D., Sr., 47, 49
Chicago (speedway), 8
Christy, Bob, 130
Chrysler, Walter Percy, 108
Clark, Rodney, 129
Clymer, Floyd, 64, 83, 92
Coffin, Harry E., 6
Comstock, Gene, 129
Conde, John, 125
Constable, Trevor, 104
Counselman, Lee, 6
Courage, W. L., 25
Cunningham, Briggs, 73, 75

Denman, Hal, 105
Derham Custom Body Company, 47, 49
Detroit Public Library, 8
Dillinger, John, 11
Dow-Jones, 106
Drive-Master transmission, 9, 35, 41, 42-43

Earhart, Amelia, 11
Earle C. Anthony, Inc., 46

Eisenhower, Dwight D., 98
Emory, Clair, 84
Essex (motorcar), 9
Estaver, George W., 25
Eubanks, Joe, 129, 130, 131
Exner, Virgil, 101

Fair, D. C., 84
'Fast mail,' 9
Federal Reserve Board, 66, 125
Ferry, Hugh, 109
Figaro, Lou, 129, 130, 131
Fisher & Ludlow, 113
Fitzpatrick, Art, 35
Fleming, George, 130, 131
Flock, Bob, 129
Flock, Fonty, 67, 70, 130
Flock, Tim, 69, 70, 71, 72, 129, 130
Folger, Guy, 59
'Ford blitz,' 88, 89, 125
Ford, Henry II, 88
Frazer, Joe, 125

Galdrisi, Bill, 130
German, Al, 10
Guardian Trust Company, 11
Gustav (King of Sweden), 24

Hadley, C.A.J., 25
Hallas, H. E., 106
Hamilton, R. N., 25
Hamilton, Red, 129
Hanks, Sam, 130
Harrison, Bill, 130
Hartmeyer, J. J., 81, 82
Hedge, Frank, 21, 29, 58, 62
Hendry, Maurice, 60, 111, 122
Hepler, Russ, 130
Hoffman, Paul, 109
Hood, A., 26, 51
Hudson-Essex-Terraplane Club, 47
Hudson, Henry, 6
Hudson, J. L., 6
Hudson Monobilt body, *see* Monobilt construction
Hudson Motor Car Company
  Defense contracts, 13, 77, 88
  Facilities
    Canada, 13, 24, 54, 56
    Great Britain, 13, 18, 24, 56
    Jefferson Avenue, 7, 8, 13, 17, 30, 33, 43, 56, 59, 77, 122
    New Castle, PA, 42, 43
  Formation of, 6

Jackson, R. W., 26, 51
Jackson, Roscoe B., 6
Jaderquist, Eugene, 66-68
Joe Weatherly Museum, 70
Johnson, Junior, 131

Kaiser, Edgar, 116
Kaiser, Henry, 15, 125
Katona, Iggy, 130
Keller, Al, 74, 130
Kibiger, Arthur, 15-17, 18, 20, 30, 32, 33, 34, 35, 36, 37, 39, 40, 44
Kirble, Irwin, 131
Kirby, Bill, 35, 36, 39
Koto, Bob, 35

Lamm, Mike, 15, 16, 81, 83, 98, 125
Le Mans circuit, 75
Letner, Danny, 129, 130
Ligouri, Ralph, 131
Linden, Andy, 130
Little, Major W. C., 13
Loewy, Raymond, 40, 103
Luptow, Frank, 129

Macauley, Alvan, 108, 109
Mantz, Johnny, 70, 130
Martin Company, 88
Mason, George, 96, 107-108, 109-111, 112, 115, 124, 125
McCahill, Tom, 75, 76
McGinnis, Pop, 130
McGrath, Jack, 129, 130
MacMinn, Strother, 35, 102
Meyers, Dick, 130
Michael, Arthur, 35-36
Mignault, Dorothy, 84
Milestone Car Society, The, 40
Miller, Al, 10
Miller, Chet, 10
Milton, Walter S., 25
Minneapolis Auto Show, 102
Minx, Walter, 129
Monobilt construction, 21, 32-33, 48
Moon, Tommy, 129
Moore, Bob, 129
Moran, Jim, 81-82
Mori-Kubo, Bruce, 9, 74, 102, 103
Mounts, Arden, 131
Mulford, Ralph, 8
Mundy, Frank, 74, 129, 130, 131
Munger, G. W., 51
Murphy Company (Walter M.), 9, 11
Murray Company, 9, 83, 87, 88
Muzzey, Elwin, 105
Myerson, Elliott, 105

Nance, James J., 46, 87, 109-110, 111, 126
Nash, Charles W., 107, 108
Nash Motor Company, 108-112
New York Stock Exchange, 111
Northrup, H. M., 11, 26, 34, 51

Olaf (Crown Prince of Sweden), 24
Osterman, Hans A. B., 24

Packard Motor Car Company, 108-111
Parks, Bill, 131
Patterson, Johnny, 130
Penn, Stanley, 106
Penobscot Building, 106
Penrose Trophy, 9, 10
Peterson, Bob, 131
Petitt, Paul, 131
Petty, Lee, 70
Pichurski, John, 106
Pittsburgh Plate Glass, 82
Potter, Dave, 118
Pratt, George H., 17, 25, 26, 49, 50, 51
Price, Hickman, Jr., 28-29
Prohibition, 9

Quotes from the Experts, 86

Races
  Basso Presto Sports Car Club Hillclimb, 74
  Carrera Panamericana, 70, 100
  Charlotte, NC, 70
  Columbia, SC, 68
  Darlington, 70
  Daytona, 57, 67, 68, 71
  Detroit 250, 69
  Gardena, CA, 68
  Gene Connors Memorial Hillclimb, 74
  Jacksonville, FL, 70, 71
  Langhorne, PA, 70, 72
  Macon, GA, 70
  Martinsville, VA, 71, 72
  Occoneechee, NC, 70
  Phoenix, 68
  Pikes Peak, 9, 10
  Southern 500, 69, 71
  West Palm Beach, 71
Racz, Victor, 81, 103, 105
Railton (motorcar), 10
Railton, Reid, 10, 76
Rathmann, Dick, 69, 71-72, 74, 129, 130, 131
Reed, Jim, 130
Rhoades, Tom, 67
Roberts, Bob, 58
Roberts, M. M., 25
Robinson, Ray, 103
Romine, Jim, 130
Romney, George, 107, 108, 109, 110, 111, 120, 122
Ruttman, Troy, 131

Safety Engineering's 'America's Safest Car' trophy, 50
Sager, Buck, 129, 130
Selective sliding gear transmission, 6
7-X engine, description of, 72-73
'Severe usage' items, listing of, 70-71, 72
Shaw, Wilbur, 77, 93-95
Sheepshead Bay Raceway, 8
Shuman, Buddy, 129
Smith, Jack, 129, 130
Smith, T. Marion, 57
Smith, Yates G., 26
'Speedway bodies,' 8
Spring, Clara, 81, 105
Spring, Frank, 11, 12-13, 15, 16, 30, 35, 37, 40, 76, 80, 81, 83, 98-99, 100, 101, 102, 103, 104, 105
Steinbronner, Fred, 130, 131
Step-down convertible, construction of, 44
Stratton, J. A., 104
Studebaker Corporation, 109-111
Supermatic drive, 53-54
Swengles, I. B., 51

Taruffi, Piero, 70
Taylor, Bill, 130
Taylor, Jesse, 69, 129
Teague, Dick, 104
Teague, Marshall, 58, 61, 63, 66, 68-69, 70, 71, 72, 73, 75, 116, 129, 130, 131
Templin, H. L., 72, 73
Terraplane, 10
Tevfik Ercan of Betkas, Ercan ve Seriki, 43
Thomas (Hudson export manager), 10
Thomas B. Jeffery Company, 108
Thomas, Don, 130, 131
Thomas, Herb, 66, 67, 68, 69-70, 71, 74, 129, 130, 131
Thompson, Jim, 129
Thompson, Tommy, 70
Tippet, Lennie, 129
Toncray, Millard H., 12, 34, 35, 83, 103
Touring, *see* Carrozzeria Touring
Treadwell, C. W., 25
True Center-Point Steering, 22
Turner, Curtis, 70
Twin H-Power, description of, 70-71

Unit engineering (unit construction), 9, 32, 33, 44
Unit power, 7
United States Government
  Defense Department, 125
  Office of Price Administration, 23
  Office of Price Stabilization, 64, 66

Vacumotive Drive, 35, 41
Vail, Ira, 8
Vance, Harold, 109
VanDerZee, N. K., 25, 61, 95, 114
Vorderman, Don, 74

Wall Street Journal, 106
Weather-Master, 27
Wheat, Fred, 130
Whittaker, Wayne, 41-43
Wilhelmina (Queen of Holland), 109
Williams, C. Mennen ('Soapy'), 49
Woron, Walt, 63-64, 77
Wright Corporation, 88

X-type frame, 22

Yoder, Ebe, 130
Yonkers, Arnold, 35
Young, W. E., 25